WINGED WORLDS

This edited collection explores our often-surprising modes of co-inhabiting the cultural and aerial worlds of birds. It focuses on our encounters with non-captive birds and the cultural geographies of feathered flight.

This book offers a timely contribution to the more-than-human geographies of flight, space and territory. The chapters support an ethics of attention as a new basis for the conservation and cultivation of aerial habitats. Contributions adopt an interdisciplinary approach to the patterns of intrusion and escape that shape our encounters with birds and unsettle our traditionally terrestrial concepts of space. Each chapter focuses on a different aspect of our shared lives with birds, ranging from scientific observation to the social media-enabled spectacle of co-habitation and spatial competition.

Written in a thought-provoking style, this book seeks to address a dearth of critical perspectives on the cultural geographies of flight and its implications for the ways in which we understand common spaces around and above us in the context of any effort at conservation.

Olga Petri is a Leverhulme Trust Early Career Researcher in the Geography Department of Cambridge University. Her main interest is in the cultural and historical geographies, social communities, and more-than-human assemblages in urban spaces shaped by the modern bureaucratic state. She is the author of *Places of Tenderness and Heat: The Queer Milieu of Fin-de-Siecle St. Petersburg* (2022).

Michael Guida is a Research Associate in Media & Cultural Studies at the University of Sussex. He is a writer and a historian of nature in modern British urban culture, with a particular interest in human–avian relations. His first book is called *Listening to British Nature: Wartime, Radio & Modern Life, 1914–1945* (2022).

Routledge Human–Animal Studies Series
Series edited by Henry Buller
Professor of Geography, University of Exeter, UK

The new *Routledge Human–Animal Studies Series* offers a much-needed forum for original, innovative and cutting-edge research and analysis to explore human–animal relations across the social sciences and humanities. Titles within the series are empirically and/or theoretically informed and explore a range of dynamic, captivating and highly relevant topics, drawing across the humanities and social sciences in an avowedly interdisciplinary perspective. This series will encourage new theoretical perspectives and highlight ground-breaking research that reflects the dynamism and vibrancy of current animal studies. The series is aimed at upper-level undergraduates, researchers and research students as well as academics and policy-makers across a wide range of social science and humanities disciplines.

Horse Breeds and Human Society
Purity, Identity and the Making of the Modern Horse
Edited by Kristen Guest and Monica Mattfeld

Immanence and the Animal
A Conceptual Inquiry
Krzysztof Skonieczny

The Imaginary of Animals
Annabelle Dufourcq

Winged Worlds
Common Spaces of Avian-Human Lives
Edited by Olga Petri and Michael Guida

Methods in Human–Animal Studies
Engaging with animals through the social sciences
Edited by Annalisa Colombino and Heide K. Bruckner

For more information about this series, please visit: www.routledge.com/Routledge-Human-Animal-Studies-Series/book-series/RASS

WINGED WORLDS

Common Spaces of
Avian-Human Lives

Edited by Olga Petri and Michael Guida

Routledge
Taylor & Francis Group

LONDON AND NEW YORK

Designed cover image: © Xavier Bou

First published 2023
by Routledge
4 Park Square, Milton Park, Abingdon, Oxon OX14 4RN

and by Routledge
605 Third Avenue, New York, NY 10158

Routledge is an imprint of the Taylor & Francis Group, an informa business

British Library Cataloguing-in-Publication Data
A catalogue record for this book is available from the British Library

ISBN: 978-1-032-36971-6 (hbk)
ISBN: 978-1-032-36972-3 (pbk)
ISBN: 978-1-003-33476-7 (ebk)

DOI: 10.4324/9781003334767

Typeset in Bembo
by SPi Technologies India Pvt Ltd (Straive)

CONTENTS

FIGURES

ACKNOWLEDGEMENTS

Olga and Michael would like to thank all contributors to this collection for their intellectual curiosity and commitment to the project. We also feel grateful to all members of the *Winged Geographies Group* and everyone who joined us for our annual workshop in 2021 to share their interest in birds and other winged creatures. Special thanks go to Philip Howell for his encouragement, support, and patience when reading drafts of the introduction to this collection.

We also would like to thank the Geography Department, University of Cambridge, and the Media and Cultural Studies Department at the University of Sussex for their institutional support. The project was generously funded by the Leverhulme Trust (Olga's ECF-2017-017).

CONTRIBUTORS

William M. Adams is a geographer based in Geneva and Cambridge. He is interested in the way novel technologies shape ideas and practices in nature conservation. His book, *Strange Natures: Conservation in the Era of Genome Editing* (with Kent Redford) was published in 2021.

Shawn Bodden is a research associate at the University of Glasgow, Scotland, UK. His research concerns the everyday politics of dissent, resistance and protest in urban contexts. His approach is influenced heavily by ethnomethodology, and he is particularly interested in how this approach to studying social interaction can be expanded to consider non-human sociality.

Philip Howell is professor of Historical Geography at the University of Cambridge. He has written monographs on the regulation of prostitution in Britain and on the British Empire and on the domestic dog in Victorian Britain. The latter reflects his longstanding interest in historical animal geographies, but he has also written on relations between humans and animals in contemporary societies.

Patricia Jäggi is a cultural anthropologist working in the field of sound studies, with a focus on listening and sound in interspecies relationships and sound ecology. Her current research focuses on sonic human–bird relations. She is also a passionate field recordist and uses audio technology to approach and mediate environmental knowledge. She works as a senior researcher at the Lucerne University of Applied Sciences and Arts – Music.

Dolly Jørgensen is professor of History at University of Stavanger, Norway. Her current research focuses on cultural histories of animal extinction. Her monograph

Recovering Lost Species in the Modern Age: Histories of Longing and Belonging was published with MIT Press in 2019. She has co-edited four volumes and is co-editor-in-chief of the journal *Environmental Humanities*.

Alex Lawrence is a doctoral researcher in Modern Languages at the University of Oxford. His work brings together multiple aspects of early modern culture, from natural history, travel narrative, and confessional writing to collecting and painting. He is passionate about the natural world: something that surfaces in his research on the Toucan, but extends in his life as a birdwatcher, walker, and wild-swimmer.

Paul Merchant is an oral historian with a background in cultural geography. His publications are concerned with the production of environmental knowledge, the history of recent science and the use of interviews in the historiography of science and technology.

Andy Morris is a senior lecturer in the department of Geography at the Open University. He comes from a background in cultural geography and has written across a range of subjects relating to landscape, visual culture, nature and human–wildlife relations. Andy's writing has been published in Open University teaching publications and peer-reviewed journals. He has also worked as an academic consultant on two BBC television series, most recently the BBC series Autumnwatch. Andy's recent work on human–starling relations has seen him carry out fieldwork in Rome.

Jeremy Mynott is the former CEO of Cambridge University Press and an emeritus fellow of Wolfson College, Cambridge. He is the author of *Birdscapes: Birds in Our Imagination and Experience* (2009), *Thucydides* (2013) and *Birds in the Ancient World: Winged Words* (2018) and co-author (with Michael McCarthy and Peter Marren) of *The Consolation of Nature* (2020). He is a founder member of 'New Networks for Nature'.

Sara Asu Schroer is a social anthropologist based at the Department of Culture Studies and Oriental Languages at the University of Oslo. Since early on birds have played a central role in her research, which contributes to cross-disciplinary debates in environmental studies (i.e. on domestication, hunting, conservation and more-than-human ethnography). She is co-editor of *Exploring Atmospheres Ethnographically* (Routledge: 2018).

Adam Searle is a cultural and environmental geographer based at the University of Nottingham. His research broadly examines the politics and relations between humans, other animals, science, and technology in the context of ecological breakdown and global environmental change. He is a co-founder of the Digital Ecologies research group and an editor of the forthcoming collection *Digital Ecologies: Mediating More-than-human Worlds* (Manchester University Press).

Jonathon Turnbull is a cultural and environmental geographer at the University of Cambridge interested in more-than-human geographies and digital ecologies. His PhD research focuses on the weird ecologies of the Chornobyl Exclusion Zone, especially dogs and wolves. He is a co-founder of the Digital Ecologies research group.

Andrew J. Whitehouse is a lecturer in the Department of Anthropology at the University of Aberdeen, UK. His research is predominantly in environmental anthropology and human–animal relations, particularly issues relating to birds and nature conservation. He worked on the AHRC-funded Listening to Birds project that explored people's relations with birds through sound and has conducted fieldwork on conservation issues on the island of Islay in the west of Scotland.

Roger S. Wotton is Emeritus Professor of Biology at UCL. In addition to research on organic matter in fresh waters, he taught courses in Aquatic Biology and in Animal Form and Function, the latter including animal locomotion. An interest in Victorian Natural History led to a double biography of Philip Henry Gosse and his son Edmund and many blog posts. In addition, Roger continues his lifelong passion for paintings and has given Lunchtime Talks, and lectures, at the National Gallery in London, a recent course being on Angels and Demons.

LEARNING TO LIVE IN WINGED WORLDS

Introduction

Olga Petri

Avian Encounters

A few years ago, I was woken one night by a raucous knocking and scraping. I turned on the light to investigate. Seeing nothing that moved, I noticed that the noise came from inside a sealed fireplace. I tore away a few loose boards to create an opening large enough to peer inside. Using the LED on my phone, I saw a pigeon flapping about in the small cavity and raising a cloud of dust and dry faeces. I had often seen pigeons sitting or nesting on the chimneys. By some rare circumstance – perhaps a particularly forceful gust of wind or a loose piece of pottery along the rim – this one had stumbled into the flue and fluttered its way down. Luckily, she fell into one of the shallowest chimneys available.[1] Having now been discovered, she redoubled her frantic flapping.

To save my involuntary guest, I tore away more boards and tried gently to lift her out. Each time I tried, she fluttered, hopped, and forcefully struck her body against the walls. After some time, having been pecked at, scratched and defaecated upon, I finally clasped her between my two slightly bloodied hands. Once the bird felt that my touch was gentle, she stopped resisting and went almost completely limp – as it turned out, not from exhaustion. I carried her outside. To my relief, she raised her wings almost immediately and flew off, disappearing into the night sky.

This incident struck me in two ways:

Firstly, it demonstrated that, bereft of the power of flight, pigeons are remarkably vulnerable and clumsy creatures, awkward and almost comically unbalanced. Flight alone lends them and many other birds their peculiar grace, as well as the ability to flee with impunity, at a moment's notice. Birds take flight but they also *make* flight, or, better still, they *are* flight: that is, they embody (even the flightless amongst them) the dimension of the air, even that mythic zone of light and fire, the empyrean. Watching birds "swoop and glide and carve their way through the

DOI: 10.4324/9781003334767-1

unseen air", writes the cultural ecologist David Abram, "surely ignited many of our most human aspirations toward freedom and flight" (2010, 273). Human beings are bound and ground to the earth, shaped by the soil, but we look up to the birds who fleetingly descend to our level. They visit our world, but they also live and move beyond us. Living with birds means becoming acutely aware of our limitations, even while birds inspire us to imagine a different way of inhabiting the space that envelops us. Abram rightly notes that "flight itself is a kind of thinking, a gliding within the mind" (2010, 269). The contributions to this collection focus on birds in various dimensions (symbolic, conceptual, and physical), considering the terrestrial and aerial spaces we – often unwittingly – share with birds, and the spaces of the human imagination, our flights of fancy. These spaces are our "winged worlds".[2]

Secondly, *my* poor pigeon brought home to me the tension between abstract theorising and the lively encounter with animals' otherness. Nothing I had read about pigeons and their place in our world had really prepared me for this meeting. I was aware, of course, of pigeons in or near human habitations as pests or "matter out of place" (Philo 1998, 52). I was also familiar with the meta-discourse that sees this characterisation as symptomatic of the ways in which we classify forms of life and seek to impose order on them, sometimes in cruel and unproductive ways, or in ways that reinforce our own sense of community (Jerolmack 2008). None of these approaches prepared me for the pigeon in my fireplace! While perhaps particularly out of place, she didn't strike me as in any way extraordinary. Obvious instead was that she was an individual in need of my help. Unlike the iconic pigeons who have made themselves at home in London's Trafalgar Square or on Brooklyn rooftops, this pigeon did not want to be in my fireplace and, at least initially, played no meaningful role in my world or challenge my identity. The chimney to her was nothing but an unintended snare, and for me, her position was a call to action. I was not defending territory, but disentangling it in a way that could not be but mutually beneficial. As my fearful curiosity gave way to the practical demands of her predicament, the pigeon's stance towards me seemed to change. Frantic resistance turned into gentle submission. I read this change as an acceptance of our new, friendlier footing, and I could not help but marvel at the uncanny intelligence that calmed the pigeon as she permitted me to carry her down two flights of stairs, down a hallway and out the front door. For the moment at least, we were engaged in a joint enterprise: the restoration of flight for her and satisfied sleep for me.

In this collection, the "encounter with otherness" is an ever-present theme (Abram 2010, 267). As in my own experience, an element of surprise is a typical characteristic, accompanied by asymmetrical risk and mutual dependence. Close communion is not always the result. Forrest Clingerman writes about herons, whose flight keeps them at what he calls an intimate distance from human observers. These birds' propensity to "migrat[e] beyond our conceptions of them, as flying outside our frames of reference" maintains a presence laden with absence (2008, 323). There are challenging encounters of this sort, sudden and slightly uncanny openings into nature, in a number of chapters in this collection, such as the search for an elusive bird on a farm survey in England (Paul Merchant), the manning of a falcon (Sara Asu Schroer), the murmuration of starlings on a tourist visit to Rome (Andy Morris), a joint commute with human and pigeon passengers on public transport in London (Shawn Bodden), a recording session with arctic terns in Iceland (Patricia Jäggi), a visit to a taxonomical display of an extinct bird (Dolly Jørgensen), and even a fantastical negotiation with mythical winged creatures in a Greek comedy (Jeremy Mynott) and the biblical annunciation (Roger Wotton). In many cases, the affective intensity of these encounters is heightened by the contrast between the human animal's relative immobility and these birds' capability of flight. Flight can end the encounter early or avoid it altogether. It is not a coincidence that many of the encounters that are so central to the chapters of this collection involve one party being where they are not commonly expected to be, such as a pigeon in a train or a researcher in a field of breeding birds. These unexpected encounters are an important reminder that our mutual responses reflect small-scale affective geographies that are not easily integrated into a consistent conceptual landscape (Wilson and Tewdwr-Jones 2020). We need to hold on to the otherness inherent in such encounters. In this sense, Shawn Bodden, a contributor to this collection, recommends "ethnomethodological studies of interaction as one way to think about and inform these debates […] through the analysis of actual encounters between humans and pigeons".

This collection takes more than a dove step in that direction, insofar as it represents, from an interdisciplinary range of perspectives, an attempt to take seriously the often surprising, often unforgettably vivid nature of our encounters with birds – and to use these flight-charged encounters as a basis for re-examining not so much the normative ethics of our co-habitation with birds, as the opportunities for a more subtle and attentive approach to the aerially inflected and more-than-human geographies they represent. In this collaborative and deliberately non-linear project, *fleeting* encounters with birds are of central importance. Flight obviously unsettles and challenges our traditionally terrestrial conceptions of place and space, and also the associated conventions of cultivation, domestication, planning, zoning, and conservation. Given its disruptive implications and its potential to move us through unexpected encounters, it is perhaps surprising that the cultural significance of the encounter with birds and their capacity for flight has received so little critical attention in animal studies, although attention to birds is developing significantly.

The academic discussion about the cultural significance of birds has focused on practices such as bird watching and conservation, on the migration, domestication, and consumption of birds, on co-habitation with urban birds, and on the history of ornithology. Exceptions, focusing decidedly on the encounter and on the role of flight in facilitating or averting it, come mainly from outside of academia and include such works as *Vesper Flights* and *H is for Hawk* by Helen MacDonald, the prose poem *Crow Country* by Mark Cocker, or Tim Dee's book *Landfill* about seagulls (Macdonald 2014, 2020; Cocker 2016; Dee 2019). Much recent critical work on birds has addressed non-captive birds, suggesting a growing appetite for books about the contributions of wild birds to human culture, the history of ornithology, and the cultural practices of bird watching, of birdscapes and soundscape created by birds (Fine and Christoforides 1991; Macdonald 2002, 2014; Dunlap 2009; Jeremy Mynott 2009; Mundy 2009, 2018; Nagy and Johnson 2013; Birkhead 2014; Cocker 2016; Cherry 2019; Leap 2019; Guida 2021; Jacobs 2021, 2016). Domesticated birds understandably play a crucial role in the more-than-human ethics and politics of food production and have also become the subjects of a growing body of academic work (Price 1998; Squier 2010; Otter 2020; Oliver 2021). Interesting here is the interaction with farming, which, until recently, has been considered more benign than may in fact be the case (Shrubb 2003; Musitelli et al. 2016; Bretagnolle, Denonfoux, and Villers 2018). Similarly, the social and cultural history of sports involving birds, including breeding and racing of pigeons, hawks, and falcons, has attracted recent attention (Johnes 2007; Jerolmack 2008; Day 2019; Nixon 2022). Others yet have written insightfully about the cultural and practical challenges of conservation and terrestrial co-habitation with birds (Čapek 2005; Lorimer 2008; Nagy and Johnson 2013; Bargheer 2018; Gandy 2022a).

This collection complements these works. It addresses a growing appetite for scholarly reflection about birds but does so from the perspective of cultural geographies of flight, of bird's often puzzling patterns of presence and absence, their potential to inform human thinking about space, time, progress, and community. Andrew Whitehouse (who contributes to this collection), with Richard Smyth and others, has already begun to open up new thinking about birds in this area, but this collection broadens the scope for interdisciplinary critical reflection upon bird flight (Van Dooren 2014; Whitehouse 2015; Smyth 2020; Rose 2021). In addressing the topic of flight-on-feather, *Winged Worlds* also engages with work on other species that puts spatial concepts front and centre. This includes, for example, histories of more-than-human colonial labour, hunting and wildlife observation in national parks, shepherding and walking with dogs, and narrative selection and editing in natural history television (Philo and Wilbert 2000; Laurier, Maze, and Lundin 2006; Kalof and Montgomery 2011; Lorimer and Srinivasan 2013; Buller 2016). Entering a conversation with these works, the contributions in this collection revisit traditional animal studies themes, such as animal agency, identity, and animalisation, but emphasise the limits of human understandings of the (to us) unreachable, untouchable, sometimes indifferent spaces that birds penetrate via their capacity for flight.

Winged Worlds: Avian–Human Relations

First, consider the connections between birds, flight, space, and coexistence with human beings. In a variation on Claude Lévi-Strauss' famous observation that "animals are good to think with", Stefan Bargheer suggests that birds "are so popular *not* because they are good to think with but because they are good to play with"(my italics) (2018, 7) The kinds of *games* he has in mind are quite varied but stand in contrast to a perception of non-human animals as simply spoils and sustenance available for plunder from nature's inexhaustible factory, or even of animals as a convenient screen for the projection of human meanings and relations. Notably, these games include ones that might be understood as such in the literal sense – for instance, the Brooklyn pigeon races Colin Jerolmack studies, or the falconry practices Sara Asu Schroer describes in her contribution to this collection.

This does not imply, of course, that human experience of non-human animals – and their scholarly treatments of the latter – have moved unambiguously in a cooperative and playful direction. Human modes of engagement with birds take many forms, and few of them feel game-like. There is the routine consumption of highly commodified and remotely produced "poultry" and other kinds of "game". There are other forms of consumption, such as hunting and bird-keeping, both central to various identity- and community-building projects (Leap 2019), as well as continuing encroachment upon managed or shared habitats (Čapek 2005; Dunlap 2009; Nagy and Johnson 2013).

At a more abstract level of engagement, there is the practice of science and natural history, where wild birds, often imagined in a rarefied version of nature as a place apart from human society, have long attracted attention. Elizabeth Cherry tells the story for instance of John James Audubon's "naturalist's gaze", seeing in this the pendant or even antidote to the urbanite's blasé attitude described by Georg Simmel (Cherry 2019). From the start, however, the naturalist's gaze is anything but unproblematic, initially articulated alongside practices of specimen collection that Dolly Jørgensen in her contribution to this collection links to the extinction of several species of birds. Similarly, conservation programmes were and often remain infused with exclusionary attitudes towards "foreign" or "invasive" species (Cherry 2019, 15).

Today, according to Spencer Schaffner and Thomas Dunlap, bird watching – and listening! – are games of sorts, which continue to attract dabblers and committed aficionados (Dunlap 2009, 9; Schaffner 2011). And yet, here also, an occupation that "began with science, conservation, and self-improvement" is today reflected in field guides, according to Schaffner, in a sanitised and misleading representation of an idealised nature (Schaffner 2011, 6). In a similar context, Mathew Gandy presents a systematic critique of a political ecology that rests upon the fetishisation of nature, which Jennifer Price describes well for the less theoretically inclined in a chapter on a store chain called Nature Company in her book *Flight Maps* (Price 1998; Gandy 2022b).

Further to such critically nuanced views of "birds in nature", the role of birds in the construction of symbolic or metaphorical linkage has been influentially described by Gary Alan Fine and Lazaros Christoforides (1991). In this collection, Philip Howell expands this discussion by looking at the language of identity construction and social or political exclusion as it connects precariously placed birds and human communities (Jerolmack 2008). Flight, here, continues to play an important role. Bernard Quetchenbach, in his contribution to a collection entitled *Trash Animals*, looks at the ever less accommodating attitudes of rural populations towards Canada Geese as these abandon their once-habitual annual migratory patterns (Quetchenbach 2013). As varied as the birds themselves, the scholarly attention committed to our encounters with them by the contributors to this collection serves as a reminder that where patterns of consumption, displacement, or aestheticised observation and metaphorical extrapolation are the norm, more than one piece of the puzzle is still missing from what we have here called "Winged Worlds".

Our title refers explicitly to the very body parts which are most instrumental to flight, which shape birds' worlds by integrating aerial and terrestrial habitats. These worlds are not entirely distinct from our own. They overlap and are, of course, significantly affected by human activity. Nevertheless, the vast majority of the three-dimensional space they encompass is well beyond the reach of unaided human motion. We chose this title to reflect a sense that our encounters with birds remind us not only of their capacity for flight but also of its broader implications. They enfold us into a world qualitatively different from our own, in which space extends vastly upwards and in which absence and escape can be as intense an experience as fleeting presence. In presenting the collection under this title, we want to emphasise that this is not our world, which birds also inhabit. Instead, in winged worlds, we are better positioned as guests, tourists, onlookers. Certain modes of study, attention, and metaphorical reflection can give us vicarious access, but these will never be our worlds, no matter how much we encroach upon or threaten them.

There is a sense, however, in which the "winged worlds" we share with birds are unique. It is difficult for any one species, including birds or humans, to lay claim to and defend aerial spaces. They are intrinsically vast and borderless and, as such, an extreme example of a common good or a "common" for short. For anyone accustomed to thinking of the air as empty and infinite, this collection, with its emphasis on flight and aerial geographies related to birds, is intended to encourage further reflection on the inhabited state of the air. Wings and lines of flight penetrate and integrate this vast aerial common that encompasses and supports us as well, but is not easily accessible or navigable for our bodies and minds. Regrettably, the borderless and vast nature of this commons, the mobility of its inhabitants, and the ephemeral or nearly intangible quality of the air all aggravate Garrett Hardin's classical tragedy of the commons and complicate the model solutions observed and so successfully publicised in Elinor Ostrom's work on human interactions with ecosystems (Hardin 1968; Ostrom 2009). We and the authors of this collection believe that our encounters with birds – with extinct species via scientific enquiry, with metaphorically anthropomorphised birds, and with living beings – create opportunities for building an awareness of the air as a shared habitat deserving of our attention and protection.

Light as a Feather: Geographies of Flight

This collection addresses the relative scarcity of work focusing on bird flight in its cultural–geographical dimensions and its challenge to the human spatial imagination. The contributors to this collection share a sense that flight and fleeing offer an opportunity to disrupt otherwise all-too-comfortable concepts of "shared" space. Our intention is to contribute to a discussion about the ways in which the physical and cultural terms of our co-habitation with birds are negotiated in very different kinds of space, ranging from terrestrial and aerial territories to the intellectual spaces of museums and animal histories, or even the digital realms of social media and surveillance. Most importantly, this collection shows that the cultural and physical presence of birds is a fundamental conceptual and practical challenge for the cultivation of territory, and the often implicit conceptual privileging of human habitation. Birds teach us "another model of how to dwell in place" (Clingerman 2008, 315). As our title indicates, this is not a single winged world (entirely out of reach for human beings) but rather a multiplicity of winged worlds that we construct, share, and negotiate with birds, even when all we have are their traces, when we lament their disappearance or seek their expulsion. From our perspective, the signal and timely value of the contributions to this collection lies in their ambition to critically revisit the dimension of flight in these winged worlds during an era (the so-called Anthropocene) when wild birds are becoming scarcer.

There are some practical challenges here that run through many of the contributions and link them to a broader set of discussions that do not lend themselves to formulaic resolution. Firstly, birds change their behaviour when they are around us (Tim Birkhead and Van Balen 2008; Tim Birkhead 2011). Some of these adjustments are observable right next to us and themselves worthy of attention, as they illustrate the range of possibilities for co-habitation and reveal the cultural interests at stake. Others require less convenient observational practices, sometimes aided by digital tracking devices, drones, miniature cameras, and microphones. Still, a scientific and ontological air of mystery clings to birds (Van Dooren 2019), which several of the contributions to this collection address directly. While a number of contributors seek ways to work around or adjust for their own intrusions or the exceptional nature of their own encounters, others instead espouse and seek to understand the mysteries that are perhaps among the reasons why birds remain such powerful metaphors and symbols. As representations of the highest human aspirations, birds' power of flight aligns them naturally with the divine and the noumenal. By combining close attention to the circumstances of avian encounters in a broad range of situations from the quotidian (commuting) to the esoteric (falconry) with a critical reading of "winged worlds" as a more-than-human linguistic landscape, the contributions in this collection offer perspectives on flight-enabled cultural and political ecologies that are uniquely sensitive to the problem of human intrusion or observation.

Secondly, birds tickle our fancy and often cause us to wax lyrical about their special powers while we conveniently forget about their vulnerabilities and thus our own responsibilities. The reality and conceptual implications of bird flight create undeniable epistemological challenges and affront us in our constrained spatialities. In the context of our vaunting ambitions and proclivity to disavow responsibilities as consumers and custodians of our planet, reflection upon winged worlds brings us face-to-face with the involuntary constraints of our ideas. The aim of this collection is to illuminate the ways in which winged worlds undermine and complement human efforts to lay claim to territory and exploit the metaphor of flight. From the ancient Egyptian ibis and Horus-falcons to Pliny's bestiaries and Leonardo da Vinci's studies of birds in flight, the avian realm and its inhabitants have served as antithesis to human forms of locomotion, interaction, and cultural construction. Often, birds are no more than a crutch for understanding the air – that critical medium or *atmosphere* so important to our lives, but also easy to overlook.

Our intellectual struggle to sense the air in which we live and its cultural association with birds is disturbingly brought home in the painting *An Experiment on a Bird in an Air Pump* by Joseph Wright of Derby, in which the air is sucked out of a glass dome while a live bird sits inside (Wotton 2020). The implication here is that we would fare no better without the air. The "unheimlich" air of the painting is addressed by Peter Sloterdijk with reference to poisoned or polluted air (Sloterdijk 2004). Here, air becomes evident when we are deprived of it, and the reactions of some of the spectators depicted by Wright affect us viscerally: we feel the bird struggling for air, and our own breath seems short and precarious. Like those with asthma and other breathing problems, we sense air, even if it invisible, when our breathing is restricted. Transported to the modern-day context, the canaries-in-the-coalmine metaphor illustrates the threat to life that comes with the effects of industrial capitalism on the air we breathe as one of the most obvious "Anthropocene" ills – particularly for the poorest and least privileged communities.

The air around and above us becomes palpable not only when it is removed or polluted. Craig Martin, who engages with a rich literature concerning "geographies of aeromobility", looks at a more mundane phenomenon: fog. Martin highlights its "immersive qualities" which facilitate spatial relations even while inhibiting visibility. "Such modalities", he explains, "hover at the juncture between presence and absence" (Martin 2011, 458). For Schroer, one of the contributors to this collection, falcons play the critical role of guide dog to this medium, helping her to become "bodily and sensually attuned" to air and weather. Martin Heidegger might call this "Stimmung", playing on the meaning of "tuning", as with a musical instrument. The geographer Jamie Lorimer describes this connection, at its best, as "affective science", in which "body, affect, and skill" combine in a "relational ontology" that supports inter-species observation (Lorimer 2008, 377, 379). So, observation-as-attunement may offer a way of living with lines of flight.

For these reasons and a host of others, any discussion about the avian realm and the role of bird flight is necessarily interdisciplinary, covering a range of modalities – from field science to cultural criticism, bio-ethics, and conservation. Still, there is a consensus among the contributors to this collection in advocating what we would describe as the "ethics of attention", where relationality and acceptable scientific, social, and ecological outcomes can be best achieved by developing and deploying skills of minimally disruptive observation. Steve Hinchliffe proposes what he calls "careful political ecologies", which are underpinned by curiosity and caution, even precaution in response to the mere possibility of co-habitation and joint becoming. He juxtaposes this to traditional scientific and conservation practices that aim at preserving specimens or reconstructing illusory timeless habitats (Hinchliffe 2008, 95). Recasting this idea of care and carefulness, Maria Paula Escobar positions pigeons in Trafalgar Square within a dialogue between more-than-human geographies and political theory. She describes pigeons as "unwelcome nature" at odds with an aspirational "imperial metropolitan identity" (Escobar 2014, 371). Humanism and civility draw their sustenance, however, from kindness and compassion as much as from hygiene and cleanliness. This discussion, therefore, bifurcates into a political reading of "out-of-place" birds (see Howell's contribution to this volume) and respectful warnings against a degree of abstraction that could distract from the actual situations in which inter-species relations are negotiated on a case-by-case basis.

Lines of Flight: Spaces of Responsibility

In her study of cattle egrets in Conway, Arkansas, Stella Čapek describes the problem of territories conceptually marked as "empty". In the case study central to her book, an egret nesting ground is razed during breeding season to make way for a housing development. As one of the inadvertent perpetrators later explained, he had thought, while sitting atop his bulldozer, that "the birds will just fly off". Čapek looks at the ways in which the community grappled with this event, variously viewed as a tragedy, a necessary evil, and a breach of animal protection legislation. One of her central undertakings is tracing the role that the abstraction of abundant "empty space" plays in justifying habitat destruction. "The cultural practice of thinking of space as empty", she writes, "is associated not only with capitalist economies but with colonialism and an imposed hierarchy of values that renders indigenous inhabitants, for example, invisible" (Čapek 2005, 208, 209). The antidote can only be a kind of nature-work that builds on lived encounters to recognise this supposedly empty space as a shared and negotiated habitat.

When looking at bird flight, the seeming "emptiness" of the air and the sky is at least as problematic as the supposed emptiness of undeveloped or unused land. Laws do a poor job of delimiting flight and the right to occupy aerial space for non-human animals. A fundamental lack of demarcation maximally distributes responsibility beyond where it might be palpably attributed and felt. And yet, this aerial habitat is more important than ever to birds as a commuting zone and avenue of escape. Access to the air is essential for birds' response to the pressing encroachment of human habitation, the demands of cultivation, and the fallout from extractive industries.

Birds' flight paths undermine traditional mappings of human territories. They are a transportation affordance within and between urban territories, cultivated landscapes, and the surfaces of our planet least touched by human activity. This network enables birds, under so much pressure in their remoter habitats, to intrude successfully into the very urban landscapes that – for the first time in history – accommodate a majority of human animals (The World Bank 2020). Within these three-dimensional landscapes, bird flight integrates dramatically different ecological habitats, such as the air, the soil, and the various bodies of water and wetlands that exist in our cities. Birds' ability to fly places them intermittently on the pavements and rooftops, in fields and in treetops, on isolated rocky cliffs and in our proudest squares, under bridges, on the canals, rivers, lakes and coastlands, in and above more distant waters – and often in the air. It is exceedingly difficult "to place them".

Flight is fundamental to the idea of being "free as a bird". Not only does it unsettle the terrestrial construction of spaces and spatial networks, but it also makes for a messy fit with the conceptual categories of pest or chattel into which we sort non-human life. After all, with the notable exceptions of chickens, turkeys, and ducks, most birds arrive in our midst on their own wings, rather than in the form of edible calories in delivery trucks. Their freedom does not end on arrival, and trying to contain them is very challenging in practical terms. Not only is the political consensus around their displacement from public spaces lacking, but the methods by which these objectives are pursued are rarely straightforward in their implementation. More commonly, displacement of birds proceeds inadvertently, rather than intentionally, and is a systematic side effect of resource-intensive living and planetary urbanisation. Birds, therefore, represent an escape from the pull of the earth and of human-dominated environments, and they resist efforts to contain them, and yet their presence amongst us remains conditional and fragile.

We can invoke the concept of "lines of flight" described by Gilles Deleuze and Félix Guattari in recognising that birds present an alternative to the stultifying geographical and spatial paradigms to which we tend to default (Gilles Deleuze and Guattari 1987; Deleuze and Guattari 1994). Our typically fleeting encounters with birds play an important role in connecting us to the polysemic implications of flight. The winged worlds to which we obtain vicarious access are determined by the asymmetric possibility of escape from every human–avian encounter: birds can usually access the "smooth space" of the air, whilst we cannot. Free birds can flee from us. "Lines of flight", in Deleuze and Guattari's usage, is translated from the French "ligne de fuite", where fuite is both flight and fleeing, ascent and escape (Massumi 1987, xvii). This kind of "flight" represents "an infinitesimal possibility of escape […] the elusive moment when change happens, […] when a threshold between two paradigms is crossed" (Fournier 2014, 121). Flight is then not just a way to get somewhere, using the air as a kind of invisible infrastructure, but also a way to ascend and get away. It is an alternative way to *inhabit* space, predicated on the ability to appear and depart unexpectedly in response to threats and opportunities (Thornton 2018).

Flight as the possibility to escape is central to Deleuze and Guattari's expansive concept of aerial habitation, and it plays a key role in the contributions in this volume. Birds' flight-enabled elusiveness informs the ways in which human and avian worlds interpenetrate, the ways in which they are co-produced and constantly re-negotiated in each encounter. The possibility of flight thus provides a framework for thinking systematically about birds and the spaces they occupy, whether they are among us or over our heads (physically and metaphorically). In this collection, lines of flight accumulate to form a nodal network of spaces spanning isolated islands, farmed fields, and commuter trains. This network is overlaid by the virtual spaces of social media, digital surveillance, and also the inheritance of human myth and metaphor. Aerial and land-based habitats, migratory patterns, and cultural allusions are crucial, alongside the shaping of space by winged animals even when their presence is less tangible and even when we try to forbid their presence.

Combining socio-cultural and historical approaches, the 12 contributions to this collection work with a range of spaces defined by physical boundaries, conceptualised via social and metaphorical practices, and integrated by literal or figurative lines of flight.

Birds challenge us uniquely to take responsibility. They prompt us not only to recognise our limits but also to recognise the air we breathe and pollute as everywhere *inhabited*, a global commons in which we only have ground-floor access. Our responsibilities vis-à-vis birds' lines of flight are different in kind to those we experience towards captive animals, companion species, and livestock, the exotic beasts we travel long and far to observe, or towards creatures of water and soil (Haraway 2008, 2003). The "winged worlds" we consider are both far and proximate, elusive and intrusive. They call for a different, diffuse notion of responsibility that embraces our fragile connections to unfamiliar geographical and conceptual spaces, even while reminding us of the limitations of human existence, the power of our systems of meaning, and the nature of our interventions. Avian lines of flight (literal ones, as opposed to Deleuze and Guattari's abstractions) provide a signal opportunity to move away from the well-established perspective of human domination towards one that emphasises our entanglements with the extensive but fleeting presence of birds.

Outline of Contributions

Winged Worlds is a collection of 12 contributions that focus on the ways in which birds inhabit space and how we encounter them. Flight plays a central role, even where the capacity for flight is conspicuous by its very absence, such as in extinct birds, or temporarily abandoned as in the case of a pigeon commuting by train. The collection is divided into three sections following three qualitatively different angles of approach to avian worlds. The first section entitled *Out of Sight, Out of Mind, and Out of Place* addresses some of the ways in which birds' ability to fly away, to escape, has made them paradoxically vulnerable. We include here the human perspective on flight, fleeing, and absence, the *s'enfuir* (to flee) element of Deleuze and Guattari's conception.

Firstly, Dolly Jørgensen examines museum exhibitions concerning local and global extinction events involving birds. According to Jørgensen, extinction understood geographically as "displacement" is necessarily a "place-specific phenomenon" involving individuals, who are the last of their species, often because the specific humans responsible for their extermination assumed that there were still other members of the same species "out there". Displays, she constitutes, are often singularly insensitive to the avian habitats of extinct birds and to the individuals whose remains are on display. Done well, however, they have a transformative power to recreate in human visitors a sense of historical ecological relations and living geographies, including those once accessible to these extinct birds by virtue of their power of flight.

In the second chapter, Philip Howell contributes an essay about changing attitudes towards birds and people who suddenly find themselves *out of place* as a result of tectonic shifts in the politics of empire and hygienic urbanity. Having migrated with the encouragement of the British government, the aptly named Windrush generation lived out their lives in often hostile policy environments. Howell draws a tentative conceptual analogy between the government-orchestrated expulsion of birds and the treatment of "unwanted" human immigrants, whose departure or return *home* might be expedited by exclusionary policies.

Flight and migration in the most literal sense are the topic of Andrew Whitehouse's chapter on the creative capacities of birds that enable airborne travel. Far from routine, many migratory birds respond to factors limiting their ability to survive in one place by flying near the limits of their physiological capacity to another place, where their chances might be better. Increasingly, these travels require improvisation and re-routing, as environmental conditions change rapidly. Whitehouse presents a perspective on migration that emphasises not only the travails of flight but also showcases the behavioural adaptations required for survival in a world shaped by climate change.

Rounding out this section, Paul Merchant looks at the reasons for inattention to birds among a population particularly exposed to them and even economically incentivised to pay attention – UK farmers. While some important circumstances are certainly biographical and individual, Merchant provides a compelling analysis of the central role of flight in British farmers' perception of birds. A sense of personal regret permeates his interviews with farmers, whose attitudes remain anything but callous, but who have begun to understand that the birds have not just left their field. They are gone for good. Some regret never having paid attention, others stopped paying attention when their livelihoods came under pressure and continue to see this prioritisation as inevitable, and others yet outsource responsibility to those involved in conservation schemes, "bird counters".

The second section, *Making Sense of Shared Space*, by way of contrast, is about presence, about sharing avian and terrestrial spaces with birds, the implications of our encounters with them, and the role of flight in upsetting our notions of belonging. This section highlights birds' non-human agency and the opportunities for more-than-human engagement.

In Chapter 5, Sara Asu Schroer describes two distinct phases of falconry practices in the UK. The first involves training flights and hunting, which challenge falconers to not only get attuned to their avian partners but also to immerse themselves in what she calls "weathering worlds" and aerial environments. As a result of this kind of engagement, for example, the wind, once understood with the help of a heightened sensitivity towards the ways in which a falcon responds to it, gradually becomes material, varied, and navigable to human participants. Similarly, she recounts and analyses her experiences observing the training of falcons in the UK, Italy, and Germany, a process called "manning", which often involves behavioural adjustments on the part of human trainers to each individual bird.

Bird sounds, apart from traditional birdsong, broadcast by stationary birds, are the focus of Patricia Jäggi's contribution about the flight noises produced by birds. Some of these are produced in the syrinx, just like bird song, but they tend to be short. Others are produced by different body parts altogether, including the wings. Jäggi describes how greater attention to these "sonations", as she calls them, can help us "resonate" with the aerial environment inhabited by most birds.

In his chapter, Andy Morris describes the divergent fates of starlings in London and in Rome. While their habitats in and around London were destroyed by processes not fully understood, starlings continue to flourish in Rome, where their collective flight and swoops, or "murmurations", have become a divisive tourist attraction.

The final contribution to this section is Shawn Bodden's critical reflection on decidedly non-academic online postings about human commuters' chance encounters with pigeons on trains. In this shared urban geography, the incursions of pigeons are not straightforwardly "problematic" or "convivial" in a consistent way, but rather complex opportunities for site- and situation-specific attention, reaction, and reasoning about the co-habitation between human and pigeon passengers. Here, infrastructure and technology become an alternative to flight, which in one of the cases described appears positively incompatible with the rules of commuting. Bodden builds on arguably reductionist readings of pigeons as "matter out of place" by Philo, Jerolmack, and Escobar, advocating renewed attention to the highly contingent mutual responses to the situated challenges of living together and, in particular, the contemporary alternatives to flight (Philo 1998, 52; Jerolmack 2008; Escobar 2014).

The third and final section in this collection is entitled *Flights of Fancy* and focuses on birds and the role of bird flight in our cultural worlds, our analogue and digital imaginations. Human aspirations in art, religion, and science to fathom or even imitate the mobility and vast habitats of birds invariably serve to emphasise human limitations. In some cases, these systematic reflections can be profoundly humbling, in others comical. A way forward may be the deployment of technology to get a bit closer than has historically been possible to the most awe-inspiring feats of birds, as well as phases and aspects of their lives that have, in the past, been hidden from view. Perhaps, we can both revel in their otherness, admire their capacity for flight, and draw on these to reinforce a sense of responsibility for our shared three-dimensional habitats.

This section is chronologically organised and begins with Greek antiquity. In Chapter 10, Jeremy Mynott compares words to wings and physical to intellectual flight in his reading of Aristophanes's "radical taxonomy" as it can be found in *The Birds*. Continuing a discussion of allegorical feathered wings, the biologist Roger Wotton reflects on the physiological trade-offs involved in avian and terrestrial life. This section showcases the impressive range of conceptual references to birds and flight, ranging from the absurd and the profound in Greek and biblical antiquity to modern-day ethical ecologies and politics of race.

Alex Lawrence skips a few centuries in his chapter and explores knowledge networks during the age of exploration as Europeans grappled with specimens and descriptions of the toucan, a bird whose range of flight meant it inhabited only the "new world" across the Atlantic. What it lacked in range, however, it soon compensated by a capacity to inspire speculation, puzzlement, and art.

Concluding this section, and the book, William Adams, Adam Searle, and Jonathon Turnbull discuss the peregrine falcon's encouraging return from the brink of extinction. In addition to recounting this recent and hopeful story, they outline new, technology-enabled modes of digital engagement between humans and birds. Here, not only does flight as a means of escape and a performance take centre stage, but actual birds, their habits, the dangers confronting them move back into focus.

Across these three sections, *Winged Worlds* covers an extensive chronological, geographical, and thematic range. The focus on flight is the thread running through these contributions. Lines of flight appear as narratives in translation, habitats in transition, urban geographies exposed to human and more-than-human forms of transgression, and, not least importantly, transcendence in the form of conceptual flights of fancy, fundamental otherness, or irretrievable absence in the form of extinction. *Winged Worlds* thus explores the ways in which birds collaborate in the making of meaning about "the world" and about the cultural geographies we share with them.

Notes

1 I do not know whether it was male or female, but I prefer this pronoun to the objectifying "it".
2 I am grateful to Jeremy Mynott for the inspiration to use "Winged Worlds" as the title, see Jeremy Mynott (2018).

References

Abram, David. 2010. "The Discourse of the Birds." *Biosemiotics* 3 (3): 263–275. https://doi.org/10.1007/s12304-010-9075-z

Bargheer, Stefan. 2018. *Moral Entanglements: Conserving Birds in Britain and Germany*. Chicago: University of Chicago Press.

Birkhead, Tim. 2011. *The Wisdom of Birds: An Illustrated History of Ornithology*. London: Bloomsbury Academic

Birkhead, Tim. 2014. *The Red Canary. The Story of the First Genetically Engineered Animal*. London: Bloomsbury Publishing.

Birkhead, Tim, and Bas Van Balen. 2008. "Bird-Keeping and the Development of Ornithological Science." *Archives of Natural History* 35 (2): 281–305. https://doi.org/10.3366/E0260954108000399

Bretagnolle, Vincent, Leopold Denonfoux, and Alexandre Villers. 2018. "Are Farming and Birds Irreconcilable? A 21-year Study of Bustard Nesting Ecology in Intensive Agroecosystems." *Biological Conservation* 228: 27–35. https://doi.org/10.1016/j.biocon.2018.09.031

Buller, Henry. 2016. "Animal Geographies III: Ethics." *Progress in Human Geography* 40 (3): 422–430. https://doi.org/10.1177/0309132515580489

Čapek, Stella. 2005. "Of Time, Space and Birds: Cattle Egrets and the Place of the Wild." In *Mad About Wildlife. Looking at Social Conflict over Wildlife*, edited by Ann Herda-Rapp and Theresa Goedeke, 195–222. Leiden and Boston: Brill.

Cherry, Elizabeth. 2019. *For the Birds: Protecting Wildlife Through the Naturalist Gaze*. New Brunswick, New Jersey: Rutgers University Press.

Clingerman, Forrest. 2008. "The Intimate Distance of Herons: Theological Travels Through Nature, Place, and Migration." *Ethics Place and Environment (Ethics, Place & Environment (Merged with Philosophy and Geography))* 11 (3): 313–325. https://doi.org/10.1080/13668790802559718

Cocker, Mark. 2016. *Crow Country*. London: Random House.

Day, Jon. 2019. *Homing: On Pigeons, Dwellings and Why We Return*. London: John Murray.

Dee, Tim. 2019. *Landfill: Notes on Gull Watching and Trash Picking in the Anthropocene*. London: Chelsea Green Publishing.

Deleuze, Gilles, and Félix Guattari. 1987. *A Thousand Plateaus: Capitalism abd Schizophrenia*. Minneapolis: University of Minnesota Press.

Deleuze, Gilles, and Felix Guattari. 1994. *What it Philosophy?* Edited by Hugh and Burchell translated by Tomlinson, Graham. Columbia: Columbia Universoty Press.

Dunlap, Thomas R. 2009. "Inventing the Birdwatching Field Guide." *Forest History Today* 54 (Spring/Fall): 48–55.

Escobar, Maria Paula. 2014. "The Power of (Dis) Placement: Pigeons and Urban Regeneration in Trafalgar Square." *Cultural Geographies* 21 (3): 363–387. https://doi.org/10.1177/1474474013500223

Fine, Gary Alan, and Lazaros Christoforides. 1991. "Dirty Birds, Filthy Immigrants, and the English Sparrow War: Metaphorical Linkage in Constructing Social Problems." *Symbolic Interaction* 14 (4): 375–393. https://doi.org/10.1525/si.1991.14.4.375

Fournier, Matt. 2014. "Lines of Flight." *TSQ: Transgender Studies Quarterly* 1 (1–2): 121–122. https://doi.org/10.1215/23289252-2399785

Gandy, Matthew. 2022a. "An Arkansas Parable for the Anthropocene." *Annals of the American Association of Geographers* 112 (2): 368–386. https://doi.org/10.1080/24694452.2021.1935692

———. 2022b. "Urban Political Ecology: A Critical Reconfiguration." *Progress in Human Geography* 46 (1): 21–43. https://doi.org/10.1177/03091325211040553

Guida, Michael. 2021. *Listening to British Nature: Wartime, Radio, and Modern Life, 1914–1945*. Oxford: Oxford University Press.

Haraway, Donna. 2003. *The Companion Species Manifesto: Dogs, People, and Significant Otherness* Chicago: Prickly Paradigm Press.

———. 2008. *When Species Meet*. Minneapolis: University of Minnisota Press.

Hardin, Garrett. 1968. "The Tragedy of the Commons: The Population Problem has no Technical Solution; it Requires a Fundamental Extension in Morality." *Science* 162 (3859): 1243–1248.

Hinchliffe, Steve. 2008. "Reconstituting Nature Conservation: Towards a Careful Political Ecology." *Geoforum* 39 (1): 88–97. https://doi.org/10.1016/j.geoforum.2006.09.007

Jacobs, Nancy J. 2016. *Birders of Africa: History of a Network*. Yale: Yale University Press.

———. 2021. "Reflection: Conviviality and Companionship: Parrots and People in the African Forests." *Environmental History* 26 (4): 647–670.

Jerolmack, C. 2008. "How Pigeons Became Rats: The Cultural-Spatial Logic of Problem Animals." *Social Problems* 55 (1): 72–94. https://doi.org/10.1525/sp.2008.55.1.72

Johnes, Martin. 2007. "Pigeon Racing and Working-Class Culture in Britain, c. 1870–1950." *Cultural and Social History* 4 (3): 361–383. https://doi.org/10.2752/147800407X219250

Kalof, Linda, and Georgina M. Montgomery. 2011. *Making Animal Meaning*. East Lansing: Michigan State University Press.

Laurier, Eric, Ramia Maze, and Johan Lundin. 2006. "Putting the Dog Back in the Park: Animal and Human Mind-in-Action." *Mind, Culture, and Activity* 13 (1): 2–24. https://doi.org/10.1207/s15327884mca1301_2

Leap, Braden T. 2019. "Gone Goose: The Remaking of an American Town in the Age of Climate Change." *Social Forces* 98 (4): 1–3. https://doi.org/10.1093/sf/soz136

Lorimer, Jamie. 2008. "Counting Corncrakes: The Affective Science of the UK Corncrake Census." *Social Studies of Science* 38 (3): 377–405. https://doi.org/10.1177/0306312707084396

Lorimer, Jamie, and Krithika Srinivasan. 2013. "Animal Geographies." In *The Wiley-Blackwell Companion to Cultural Geography*, edited by Nuala C. Johnson, Richard H. Schein, and Jamie Winders, 332–342. Oxford: Wiley-Blackwell.

Macdonald, Helen. 2002. "'What Makes You a Scientist is the Way You Look at Things': Ornithology and the Observer 1930–1955." *Studies in History and Philosophy of Science Part C: Studies in History and Philosophy of Biological and Biomedical Sciences* 33 (1): 53–77. https://doi.org/10.1016/S1369-8486(01)00034-6

———. 2014. *H is for Hawk*. London: Random House.

———. 2020. *Vesper Flights*. London: Grove Press.

Martin, Craig. 2011. "Fog-Bound: Aerial Space and the Elemental Entanglements of Body-with-World." *Environment and Planning D: Society and Space* 29 (3): 454–468. https://doi.org/10.1068/d10609

Massumi, Brain. 1987. "Notes on the Translation and Acknowledgements." In *A Thousand Plateaus: Capitalism and Schizophrenia*, edited by Deleuze, Gilles, and Félix Guattari, xvi–xx. Minneapolis: University of Minnesota.

Mundy, Rachel. 2009. "Birdsong and the Image of Evolution." *Society & Animals* 17 (3): 206–223. https://doi.org/10.1163/156853009X445389

———. 2018. *Animal Musicalities: Birds, Beasts, and Evolutionary Listening*. Middletown: Wesleyan University Press.

Musitelli, Federica, Andrea Romano, Anders Pape Møller, and Roberto Ambrosini. 2016. "Effects of Livestock Farming on Birds of Rural Areas in Europe." *Biodiversity and Conservation* 25 (4): 615–631. https://doi.org/10.1007/s10531-016-1087-9

Mynott, Jeremy 2009. *Birdscapes: Birds in our Imagination and Experience*. Princeton: Princeton University Press.

Mynott, Jeremy. 2018. *Birds in the Ancient World: Winged Words*. Oxford: Oxford University Press.

Nagy, Kelsi, and Phillip David Johnson, eds. 2013. *Trash Animals: How We Live with Nature's Filthy, Feral, Invasive, and Unwanted Species*. University of Minnesota Press.

Nixon, Sean. 2022. *Passions for Birds: Science, Sentiment, and Sport*. Montreal, Ontario: McGill-Queen's Press-MQUP.

Oliver, Catherine. 2021. *Veganism, Archives, and Animals: Geographies of a Multispecies World*. London: Routledge.

Ostrom, Elinor. 2009. "A General Framework for Analyzing Sustainability of Social-Ecological Systems." *Science* 325 (5939): 419–422. https://doi.org/10.1126/science.1172133

Otter, Chris. 2020. *Diet for a Large Planet: Industrial Britain, Food Systems, and World Ecology*. Chicago: University of Chicago Press.

Philo, Chris. 1998. "Animals, Geography, and the City: Notes on Inclusions and Exclusions." In *Animal Geographies: Place, Politics and Identity in the Nature-Culture Borderlands*, edited by Jennifer Wolch and Jody Emel, 51–71. London: Verso.

Philo, Chris, and Chris Wilbert, eds. 2000. *Animal Spaces, Beastly Places: New Geographies of Human-Animal Relations*. London: Routledge.

Price, Jennifer Jaye. 1998. *Flight Maps: Encounters with Nature in Modern American Culture*. Yale, New Haven: Yale University Press.

Quetchenbach, Bernard. 2013. "Canadas: From Conservation Success to Flying Carp." In *In Trash Animals: How We Live with Nature's Filthy, Feral, Invasive, and Unwanted Species*, edited by Kelsi Nagy and Phillip David Johnson, 153–170. Minneapolis: University of Minnesota Press.

Rose, Deborah Bird. 2021. *Shimmer: Flying Fox Exuberance in Worlds of Peril*. Edinburgh: Edinburgh University Press.

Schaffner, Spencer. 2011. *Binocular Vision: The Politics of Representation in Birdwatching Field Guides*. Boston: University of Massachusetts Press.

Shrubb, Michael. 2003. *Birds, Scythes and Combines: A History of Birds and Agricultural Change*. Cambridge: Cambridge University Press.

Sloterdijk, Peter. 2004. *Plurale Sphärologie: Schäume*. Vol. 3. Frankfurt am Main: Suhrkamp.

Smyth, Richard. 2020. *An Indifference of Birds*. Devon: Uniformbooks.

Squier, Susan M. 2010. *Poultry Science, Chicken Culture: A Partial Alphabet*. New Brunswick: Rutgers University Press.

The World Bank. 2020. "Urban Development." *The World Bank*. Accessed 10.06.2022.

Thornton, Edward. 2018. "On Lines of Flight: A Study of Deleuze and Guattari's Concept." PhD in Philosophy, Royal Holloway, University of London.

Van Dooren, Thom. 2014. *Flight Ways. Life and Loss at teh Edge of Extinction*. New York: Columbia University Press.

———. 2019. *The Wake of Crows: Living and Dying in Shared Worlds*. New York: Columbia University Press.

Whitehouse, Andrew. 2015. "Listening to Birds in the Anthropocene: The Anxious Semiotics of Sound in a Human-Dominated World." *Environmental Humanities* 6 (1): 53–71. https://doi.org/10.1215/22011919-3615898

Wilson, Alexander, and Mark Tewdwr-Jones. 2020. ."Let's Draw and Talk about Urban Change: Deploying Digital Technology to Encourage Citizen Participation in Urban Planning." *Environment and Planning B: Urban Analytics and City Science* 47 (9): 1588–1604.

Wotton, Roger. 2020. "A Bird in the Air Pump." *Winged Geographies Blog Series* (blog). 15.05.

PART I

Out of sight, out of mind, and out of place

1

DISPLAYING DISPLACEMENT

Exhibiting extinct birds in natural history museums

Dolly Jørgensen

Introduction

We are currently living through the sixth mass extinction of species on planet Earth. As humans have become aware of the extinction or imminent end of non-human animal species over the last 200 years, there have been active attempts to understand the loss (Barrow 2009; Jørgensen 2019). One place in which a reckoning with extinction happens is the museum (Guasco 2021; Jørgensen et al. 2022). Natural history museums, as repositories of animal remains (particularly as the keepers of the rarest ones) which are put on display for visitor education and edification, are one of the few contact points with extinct animals.

In this chapter, I examine, in turn, species extinction and museum exhibition as displacements. I argue that both displace animals from lively relationships, from history, and from place. Then, I examine the intersection of the two in displays of extinct birds. I discuss the displacement of these extinct birds from their ecosystems through their emplacement in the natural history museum. In this investigation of the geography of birds which can never again fly free, I probe the disconnect that is created in exhibition practice that favors isolating the extinct specimen as an aesthetic object rather than a once-living being that inhabited the world.

Extinction as displacement

Over the last 500 years, a time in which humans have created robust global capitalist networks and modified not only local ecosystems but planetary systems, human activities have created the conditions that have led to rapid extinctions and reductions in plant and animal populations. Scientific consensus is that we are living through a mass extinction event, with current rates of extinction at least 100 times

DOI: 10.4324/9781003334767-3

natural background levels. While humans have a long history of bringing about extinction as they moved around the globe from at least the Late Pleistocene (Braje and Erlandson 2013), modern extinction, termed *Anthropocene defaunation* by Dirzo et al. (2014), is pervasive: over 320 known vertebrate species have become extinct since 1500, and the remaining species have an average 25% decline in numbers. The leading global species database currently lists 159 birds that have become extinct in the last few hundred years, from the Cuban macaw to the Reunion rail to the Kauai o'o' (IUCN 2022). More birds than any other vertebrates are known to have gone extinct in the modern era. A dramatic shrinkage of vertebrate populations is happening even with species that are classified as "least concern", meaning that the extinction crisis is probably even worse than it is portrayed to be (Ceballos et al. 2017). Yet, extinction is also not an evenly distributed event—species in isolated ecosystems such as islands are easily eradicated, as are species in areas that settler colonial and capitalist practices target for so-called development—and it is not inevitable (Theriault and Mitchell 2020).

The extinction of a given species cannot be thought of as a single isolated event. Each species is intricately woven into an ecosystem with other species up and down the food chain; plant–bird interactions such as pollination and seed dispersal as well as other ecosystem functions such as decomposition and erosion control are all disrupted with extinction (Dirzo et al. 2014). Decreases in biological diversity significantly alter the biogeochemical and dynamic properties of ecosystems because everything is in a network (Naeem et al. 2021).

We should not, however, think of the connections severed by extinction only in terms dictated by Western science. Kinship, which "situates us here on Earth, and asserts that we are not alone in time or place…and we are members of entangled generations of Earth life, generations that succeed each other in time and place" (Rose 2011, 50), has been a non-Western, indigenous way of understanding ecological connection outside of the biodiversity paradigm (Whyte 2021). Kinship relations are mutual bonds that one has with another, whether those are humans, animals, plants, rocks, waterways, or more. A *kincentric ecology* understands that complex interactions between all kin (both humans and others) enhance and preserve ecosystems (Salmón 2000). Human–animal relations understood in this way are sites of engagement (Todd 2014). Whyte (2021) has argued that reciprocity through kinship is one of the key elements of the indigenous understanding of the environment because the quality of reciprocity across members of an ecosystem is what makes it thrive.

More-than-human kinship is a *lively* relation. I use *lively* here building upon Thom van Dooren and Deborah Bird Rose's concept of "lively ethographies" as "the ongoing weaving or braiding of stories" that contributes to "the becoming our shared world" (van Dooren and Rose 2016, 85 and 89) and Alice Would's quest "to find the lively dead animal within" exhibition taxidermy (Would 2021, 36). To be lively is to have agency as matter (e.g. Bennet 2010) and agency as story. Lively relations are shared connections, ones that may exist even after biological death. Liveliness is affirmed through engagement and stories of kinship.

Extinction induced by human actions, particularly as they occur within the capitalistic and colonial context, is an act of violence to these reciprocal relations of co-nourishment and co-becoming (Hernández et al. 2021) and creates environmental injustices through its assault on genealogical relations (Whyte 2021). Relations in both time and space are altered—and can be severed—through extinction (Rose et al. 2017).

Extinction is characterized by absence, with ecological niches left empty (Garlick and Symons 2020); yet, extinction is notoriously difficult to confirm because the presence of an absence is not necessarily equivalent to the absence of presence (Jørgensen 2017). It can be that an animal is still "there" without being known. In some ontologies, absence is not interpreted as passive on the part of the animals; instead, more-than-humans are ascribed agency to withdraw from the reciprocal kin relations which leads to a geographical absence (Mitchell 2020).[1] Indigenous kin relations can also continue to exist even when a species is extinct according to Western science through the integration of non-human kin into stories (Mitchell 2020).

In addition to being relational, extinction is a place-specific phenomenon. It is common in biological sciences to talk about a species being "extinct in the wild" or extinct in a particular country or extinct in a particular former range (the latter is generally called extirpation).[2] These are all markers that show that extinction is not a universal characteristic, but can have meaning based on presence or absence in a place. Symons and Garlick have urged a "place-specific attention to extinction processes, how they are shaped by geographical forces and how these processes in turn shape geographies" (2020, 292). Spatial accounts of extinction stress that contextual place-specific politics, economies, and cultures must be coupled with global geopolitics and economic structures to understand the co-production of extinction processes (Garlick and Symons 2020).

Species and places are often linked in naming practices. Take the bird we know in English as great auk, in scientific literature as *Pinguinus impennis*, and in Icelandic as geirfugl. On a sixteenth-century map of Iceland by Abraham Ortelius, a group of small islands off the southwest coast are labeled as "geie eiar" (geir islands) and one larger one as "geie fuelasker" (geirfugl rock).[3] The last large colony of the birds survived on the island known as Geirfuglasker until 1830 when a volcanic eruption forced them to move to nearby Eldey (fire island), which is also on the map. The extinction of the geirfugl in one way makes the names of these islands obsolete; in another, the extinction demands that the names remain as an inscription. The islands themselves could be like other monuments to extinction that "reveal an awareness of human–nonhuman entanglement and the wish to engage with it" (Jørgensen 2018, 197). These names are remembered on the National Land Survey of Iceland, which records these islands, but only Eldey gets a label on Google Maps. The geirfugl which made the name meaningful are gone, so the name is perhaps fading.

Or consider the Labrador duck (*Camptorhynchus labradorius*). While it wintered along the eastern US coast, this North American native bird was named after its breeding grounds in Labrador, Canada, on the far northeastern edge of the continent. The last confirmed spotting of a live individual was not, however, in Labrador but in its wintering ground in New York State in 1878. The Lost Bird Project, which has created a series of memorial sculptures by artist Todd McGrain, placed the Labrador duck sculpture in Elmira, New York, to commemorate this last sighting.[4] We know very little about the duck's breeding habits in Labrador and how it was integrated into the geographies of Canada. While the name remains, its geographical ties have been loosened.

Extinction, then, is a displacement. It is a displacement from a particular place in which something used to be found. It is also a displacement from the lively relations between humans and non-humans. This displacement always has a very specific historical progression linked to both time and space.

Exhibition as displacement

The second kind of displacement I want to discuss is that of the natural history exhibition. Natural history museums have roots in late medieval and early modern cabinets of curiosities, the Wunderkammer. The objects in these cabinets were intended to astonish their viewers—they were strange, rare, and wonderous (Findlen 1989; Daston and Park 2001). These seemingly eccentric collections chosen for the rarity or exceptionality served to raise the status of the owner, so they were often displayed in opulent settings as a sign of wealth and prestige (Findlen 1989). As a scientific approach to collecting nature took hold, the animals and plants previously displayed in the Wunderkammer were shifted to natural history museums. Some of the earliest of these were the British Museum's Natural History Museum which acquired its foundational collection in 1756 and the Muséum national d'Histoire naturelle (MNHN) in Paris which was established in 1793. By the end of the 1800s, natural history museums were being established in many major cities with the express intent of educating the public about the science of nature.

By the late 1800s, natural history museum institutions were actively collecting specimens for exhibition. Museums sent collecting expeditions to far-flung places around the globe to collect rare and usual animals, most often charismatic mammals and birds. In cases where an animal was known to be rapidly decreasing in numbers, these expeditions were intense operations to collect as many specimens as possible. These scientific collecting expeditions, such as the 1880s hunts for American bison (Shell 2004) and Caribbean monk seal (Jørgensen 2021), were framed as ventures to preserve the animals through collecting them as memories of nature lost. Field Museum geologist Oliver C. Farrington writing in 1915 argued that natural history museum collections of soon-to-be extinct nature was one of its primary values:

> A large share of the animals and plants inhabiting this continent at the time of its discovery by Europeans are not destined to survive long except as they are protected by man, and some will become extinct in spite of him. The wild pigeon, so common in Audubon's time that he saw shiploads which had been

caught up the Hudson for sale on the wharves of New York for a cent apiece, has become entirely extinct. Other birds, flowers and even minerals have also become extinct in this country since the first coming of the Europeans. To museums must be largely assigned the work of conserving the remains of such forms ere they are absolutely lost. Specimens which are valuable now will be priceless in years to come.

(Farrington 1915, 208)

The priceless-ness of extinct specimens could actually come with a cost to the animal species. The history of scientific collection's contribution to the great auk's extinction is one of best known (Fuller 2003; Kalshoven 2018; Pálsson 2020). After the colony of birds had been displaced from Geirfulgasker to Eldey, these exceedingly rare birds were feverishly hunted to provide museum specimens. As Jamie Lorimer noted, "The remaining birds and their eggs became a form of exotic animal capital, commodified and highly valued in the burgeoning market for specimens powered by museums and private collectors" (2014, 200). The end of the species is linked directly to these collecting processes. The Institut royal des Sciences naturelles de Belgique in Brussels claims that the skin of the great auk in their collection is the very last one known alive (Kalshoven 2018), and Statens Naturhistoriske Museum in Copenhagen has the organs of the last pair preserved and displayed in spirit jars. These remains are reminders that museums directly contributed to the end of the great auk and its displacement from Eldey.

When collected for public display, specimens like the great auk were often prepared as taxidermy mounts, that is, they are stuffed and positioned to appear "lifelike". Rachel Poliquin has remarked that "Taxidermy's excess of significance originates in the relationship between an original and re-animated liveliness: at once lifelike yet dead, both a human-made *representation* of a species and a *presentation* of a particular animal's skin" (Poliquin 2008, 127). Taxidermy is a craft process that removes the organs which would rot, removes the skeleton, and treats the skin with preservatives. For scientific purposes, study skins, without being on an articulated skeletal frame, are often enough. But for public display (and for some scientific purposes such as studying body conformation), an animal that looks like an animal is desired. In this case, the taxidermist will create a skeletal mount (often out of non-biological material such as wire), stretch the skin over the mount, and then stuff the body with filling material such as cotton to mimic muscles and body fat. These animal bodies have been displaced from their original habitats and reconstituted to fit into the museum habitat.

Despite that every specimen in a museum has a collection history, most natural history museums present natural history as nature without history. Specimens are on display as timeless representatives of their type. This is apparent on labels that the public reads, which tend to include biological information for the species according to the Linnean system, habitat requirements, breeding traits, size, lifespan, etc. Very rarely is the individual on display given a history. Exceptions are for individuals with anthropomorphic names who lived in captivity such as Alfred the gorilla who lived in Bristol Zoo or when the hunter who killed the animal is discussed on the label such as the tiger killed by King George V on display in the Royal Albert Memorial Museum & Art Gallery in Exeter. Excluding these few exceptions, specimens displayed in natural history museums are typically displaced in time with truncated histories.

Early natural history displays focused on Linnean taxonomy, arranging specimens by order and genus. This kind of ordering leads to birds, regardless of where they are from in the world, being grouped together in a series of shelves and glass cases. Taxidermist Charles Willson Peale had followed this trend in his Philadelphia Museum, which he founded in the 1780s. In many modern museums, this ordering preference still holds true. The extensive collection of birds on display at Naturhistorisches Museum Wien (Vienna, Austria) and Naturmuseum Senckenberg (Frankfurt, Germany) are examples of birds grouped in this fashion. These displays displace the animal from vibrant nature to the sterile case.

An alternative display technique developed in the diorama, which is a three-dimensional display which places the animal in natural foreground settings with a painted backdrop landscape (Wonders 2003). Peale pioneered the first of these bird dioramas (Andrei 2020). About this choice, he wrote, "It is not only pleasing to see a sketch of a Landscape, in some instances the habits of the animal may also be given; by shewing the nest, hollow, cave or the particular view from whence the[y] came" (Andrei 2020, 5). One of his early scenes was a bald eagle preying on a smaller songbird. By the late 1800s, habitat groups were in vogue, sometimes as a way of engaging the viewer in dramatic action scenes (like Jules Verreaux's "Arab courier attacked by lions" displayed at the American Museum of Natural History [AMNH]) and at other times giving insights into group interactions (such as William Hornaday's "The orang utan at home" also at AMNH) (Andrei 2020, 27 and 36).

Karen Wonders (2003) has shown that habitat dioramas were culturally significant, establishing European ideas of nativeness and species belonging to particular landscapes. Her study of the Biologiska museet in Stockholm shows the blending of natural and national heritage through choices made about the landscape panorama and the specimens displayed in the diorama. Without discounting her critique of the ties between biological nativeness and political meaning in the diorama, I argue on the positive side that the diorama's placement of the animal into a landscape gives them a geography. The artificial replacement habitat evokes the place that individuals of their type lived before, even though this vision is often idealized. Although a specific individual displayed may not have always come from the landscape depicted in the diorama (this is especially true with exotic specimens that may have been in zoos and those that have a large variety of habitats), the dioramas make visible that individual species live in connection with landscapes and other creatures. Diorama artists try to replicate the ecosystem based on the best available science, stressing the connections between elements in the diorama (Delahanty 2019). Animals in the diorama are emplaced in replacement ecosystems.

Collecting rare species involved killing individuals (sometimes en masse), processing the bodies through taxidermy preparation to slow their decomposition, and relocating the bodies to museum institutions for either their scientific study collections or for public display. This removal and relocation can be seen as a displacement: the individuals were displaced from the place in which they lived to a new place for their afterlife as a museum specimen. In the diorama, some taxidermied specimens are emplaced in a semblance of their original geography. Yet, even in these cases, most taxidermied individuals are displaced out of time. They are not afforded life histories in the natural history museum. They are labeled as representatives of their species rather than as individuals who lived and died and now have an afterlife on display.

The double displacement of exhibiting extinct birds

The exhibition of extinct birds sits at the intersection of the displacements of extinction and exhibition discussed above. Because the remains of extinct species are generally encountered only in the museum setting, this intersection of displacements is worth considering. In this section, I explore the junction of these displacements in the displays of extinct birds seen during my visits to natural history museums around the world, with particular reference to museums in Europe and North America.

There is a displacement from lively relations when extinct birds are displayed as representative individuals of species. Many museums put one representative specimen of an extinct bird on display. For example, at the Naturhistorisches Museum Wien and Naturmuseum Senckenberg, single exemplars of each bird species appear side by side in the cases of the bird gallery. Extinct birds here are presented to visitors mixed with and undifferentiated from all the other birds. In Vienna, the extinct birds are given one layer of differentiation from the others: they are placed in secondary hermetically sealed boxes intended to protect the valuable specimens from degradation. The grouping of specimens into the category "bird" with the secondary grouping of "order" of bird (i.e. falcons together, landfowl together, etc.) displaces the birds from any sense of the ecological relations they had when alive. The lively relations between the birds and humans are also severed in this kind of display: past human interactions with the extinct bird—whether that was through hunting, pet-keeping, religious meaning, use in clothing, or similar—are typically not put on display. The display aesthetics stresses the evolutionary relations instead. These practices separate the bodies as individual artifacts to be preserved, unconnected to the environment.

In scenarios with single representative specimens, the bird on display is almost always a male, since males tend to have more dramatic feathering. The preference for displaying males is a known bias in natural history displays (Manchin 2008). The extinct huia (*Heteralocha acutirostris*) stands out as a common exception to the single individual display because it exhibits radical sexual dimorphism, that is, the males and females have very different conformations of the bill. Displaying a pair of mounted huias to show off the female's extremely long curved beak in contrast to the male's short and pointy one hints toward the lively relations of pair bonding. The huia pair is often positioned as if they are interacting, such as the pair in the

National Museum of Scotland in Edinburgh which sit on adjacent branches of the same tree. Oral Maori accounts of huia behavior (see Monson 2005) appear to confirm strong huia pair bonds.

Many times, extinct birds are exhibited in a single case together with other extinct birds from across the globe. This connection is made explicit to the museum visitor with labeling stressing the extinct status, such as at the Manchester Museum and Naturhistoriska riksmuseet in Stockholm. These groupings ignore the bird's original geography, instead focusing on extinction as a commonality. I have argued elsewhere that this choice creates an extinction group portrait unattached to geography (Jørgensen 2022). These birds have been displaced and new categorical relations forged through the display practice which stresses the ultimate end of the species rather than its living past.

The geography of extinct birds is more radically erased in the bird gallery of the Naturmuseum Senckenberg. On their labels, the geographical range of each bird is shown using a global map and color to indicate where the bird lives. For the extinct birds, there is no range given—instead, a big red X appears over the map. While this is a dramatic way to highlight that the bird now has no range, it also erases its previous range. Because neither the map nor the textual label indicates a geography (not even a continent), and because the ordering scheme is taxonomic rather than geographical, the visitor has no geographical information of any kind about the bird's former life.

In general, animals in the natural history museum are not displayed with species histories, but when the animal has become rare or extinct, the demise of the species is often recounted for visitors with a short extinction history. The displays in the Hall of Threatened and Extinct Animals in the MNHN in Paris exemplify this. Each specimen, like the New Zealand quail (*Coturnix novaezelandiae*) and Hawaiian moho (*Moho nobilis*), is displayed individually (i.e. only one of each animal) with a text that says when the extinction happened, and sometimes, a short description of the causes.

There is a special case when a particular known individual is involved in the extinction story. Martha the passenger pigeon (*Ectopistes migratorius*), who died in 1914, was the last known passenger pigeon. Her body is in the collection of the Smithsonian National Museum of Natural History (Washington DC, USA), and whenever it is displayed, the remains are narrated with the history of her death in the Cincinnati zoo and its relationship to the death of the passenger pigeon as a species (Jørgensen and Gladstone 2022). She comes to stand in for all passenger pigeons. In fact, her history is repeated across the globe when passenger pigeons are on display rather than the history of the actual individuals in the case. The naming of this particular passenger pigeon—Martha—with a known date of death makes the story easy to narrate since it has an individual touch. Martha's death comes to symbolize the death of passenger pigeons writ large. This kind of individualization and anthropomorphizing in the extinction narrative is rather unique—the extinction of other birds are not typically narrated around the life of the last individual even when we know when and where the last one died, such as the last Carolina parakeet who died in 1917 in the same Cincinnati zoo where Martha perished (Jørgensen 2017).

Birds are generally displayed in cases with plain backgrounds, from the classic solid black in MNHN Paris to a teal color in the Naturhistoriska riksmuseet of Stockholm. These cases make it easy to observe details of the birds, but they separate the bird from any sense of a habitat. In the extinction hall of the National Museum of Scotland, there is some semblance of a background scene behind the birds on display, but it is a generic landscape not specific to the species on display.

There are, however, a few cases in which specimens of extinct birds have been set into habitat dioramas, giving a sense of place. It is not uncommon for taxidermied birds to be displayed on branches or within trees in a natural history case, and extinct birds have also been displayed this way. The passenger pigeon group in the AMNH in New York, for example, is a free-standing case with a tree branch inhabited by birds as well as several pigeons picking at seeds on the ground below. The Field Museum in Chicago has a similar smaller display with the passenger pigeons on an oak branch. Placing the birds on branches does give a sense of the environments in which the bird lived, although there is still a lack of the overall landscape.

The Labrador ducks at the AMNH are a little closer to being in a landscape, with a small display of the ducks swimming in an icy pond and walking on the snow (Figure 1.1). A small background painting behind the ducks gives a sense

FIGURE 1.1 Labrador duck diorama at the American Museum of Natural History, New York, USA. Author photograph, 2021.

of the winter landscape, but the scene is not complete—the diorama does not extend across the view, so the visitor knows immediately that it is a small painting hanging on the back wall rather than creating an immersive view as dioramas are designed to.

This can be contrasted with the only complete diorama I have found for an extinct bird: the great auk diorama at the Naturhistorisk museum in Oslo, Norway (Figure 1.2). It strives against the double displacement of extinction in museums. In this diorama, a great auk stands on a rocky island at sunset. A string of rocky islands rise from the sea. Nineteenth-century artists such as John James Audubon, John Gould, and John Gerrard Keulemans drew great auk island landscapes this way. We also notice in the diorama that the bird is not alone—behind are pairs of auks on the rocks as well as auks swimming near the shore. While the auks are bound to the land and sea, gulls and even a sea eagle soar overhead. The rock on which the bird stands is stained with bird guano; grass tufts poke through the rocks and lichen add splashes of color to the gray shapes. The diorama places this bird body into relations with geology, vegetation, and bird life. This is a lively emplacement of extinction. At a glance, it gives us a sense of how the great auk might have lived its life. The display does not negate the displacement of the extinction event—that

FIGURE 1.2   Great auk diorama at the Naturhistorisk Museum, Oslo, Norway. Author photograph, 2019.

remains a tragedy—and it is acknowledged in the signage and historical information that accompanies the diorama.

Yet, there is an unsettling detail unknown to the average museum visitor: this great auk is not a taxidermy specimen—it is a model. While the museum does hold the only great auk specimen in Norway, that irreplaceable body is kept safe in the vault. For display purposes, a modern great auk model has been created with no great auk components. Even in this unique instance of an extinct bird diorama, the physical remains of the bird itself are displaced into storage.

Countering the double displacement

The bodily remains of the extinct birds discussed in this chapter were once-living beings that inhabited the world and now inhabit only museum displays. Their displacement through extinction itself cannot be addressed, but the further displacement of these species from their relations, histories, and environment within the confines of the museum can be. I see possibilities to counter the displacement through stories and representation of place such as dioramas. There is potential to display extinct birds in ways that affirms them as having inhabited geographies that mattered to them and that they mattered to those geographies. While a diorama may only be a semblance of a geography, it has the potential to bring into light the relations between the animal and its former environment for the visitor. Even though the remains of extinct animals are treasures to be taken care of in the museum, they also are remains of relations. Exhibition practices should take care to highlight those relations. In this, there is an opportunity to allow these taxidermied dead to be lively again—to be connected to stories of life and death, of place and time, and of relations.

Acknowledgements

This research was funded by the Research Council of Norway to the project "Beyond Dodos and Dinosaurs: Displaying Extinction and Recovery in Museums" (project 283523).

Notes

1 This point was also stressed by participants in the panel "Native Responses to the Extinction of Animals" at the International Society for the Study of Religion, Nature, and Culture 2021 conference which included Andwa Nation presenters Belgica Dagua, Delicia Dagua, and Elodia Dagua who told stories of animal departures in the Amazon due to breaking of kinship responsibilities.
2 Although Garlick and Symons are critical of *extirpation* because they feel its differentiation from *extinction* renders life "fungible and exchangeable across its dynamic spatiotemporalities" (306), I think it actually adds to the geographical dimension of the discussion because it specifically calls out a species as extinct from a particular place.
3 https://www.atlasobscura.com/articles/islandia-map-sea-monsters
4 http://www.lostbird.org/project/the-lost-bird-project/

References

Andrei, Mary Anne. 2020. *Nature's Mirror: How Taxidermists Shaped America's Natural History Museums and Saved Endangered Species*. Chicago: University of Chicago Press.

Barrow, Mark V. Jr. 2009. *Nature's Ghosts: Confronting Extinction from the Age of Jefferson to the Age of Ecology*. Chicago: University of Chicago Press.

Bennet, Jane. 2010. *Vibrant Matter: A Political Ecology of Things*. Durham: Duke University Press.

Braje, Todd J. and Jon M. Erlandson. 2013. "Human Acceleration of Animal and Plant Extinctions: A Late Pleistocene, Holocene, and Anthropocene Continuum." *Anthropocene* 4: 14–23. https://doi.org/10.1016/j.ancene.2013.08.003

Ceballos, Gerardo, Paul Ehrlick, and Rodolfo Dirzo. 2017. "Biological Annihilation via the Ongoing Sixth Mass Extinction Signaled by Vertebrate Population Losses and Declines." *PNAS* 114: E6089–E6096. https://doi.org/10.1073/pnas.170494911

Daston, Lorraine and Katharine Park. 2001. *Wonders and the Order of Nature New York*. Princeton, NJ: Princeton University Press.

Delahanty, Aaron. 2019. "Six Philosophies for a Habitat Diorama Artist." *Antennae* 49: 38–47.

Dirzo, Rodolfo, Hillary S. Young, Mauro Galetti, Gerardo Ceballos, Nick J. B. Isaac, and Ben Collen. 2014. "Defaunation in the Anthropocene." *Science* 345: 401–406. https://doi.org/10.1126/science.1251817

Farrington, Oliver C. 1915. "The Rise of Natural History Museums." *Science* 42, no. 1076: 197–208.

Findlen, Paula. 1989. "The Museum: Its Classical Etymology and Renaissance Genealogy." *Journal of the History of Collections* 1, no. 1: 59–78.

Fuller, Errol. 2003. *The Great Auk: The Extinction of the Original Penguin*. Charlestown, MA: Bunker Hill Publishing.

Garlick, Ben and Kate Symons. 2020. Geographies of Extinction: Exploring the Spatiotemporal Relations of Species Death. *Environmental Humanities* 12, no. 1: 296–320. doi: https://doi.org/10.1215/22011919-8142374

Guasco, Anna. 2021. "'As Dead as a Dodo': Extinction Narratives and Multispecies Justice in the Museum." *Environment and Planning E: Nature and Space* 4, no. 3: 1055–1076. https://doi.org/10.1177/2514848620945310

Hernández, K.J., June M. Rubis, Noah Theriault, Zoe Todd, Audra Mitchell, Bawaka Country, Laklak Burarrwanga, Ritjilili Ganambarr, Merrkiyawuy Ganambarr-Stubbs, Banbapuy Ganambarr, Djawundil Maymuru, Sandie Suchet-Pearson, Kate Lloyd, and Sarah Wright. 2021. "The Creatures Collective: Manifestings". *Environment & Planning E: Nature and Space* 4, no. 3: 838–863. https://doi.org/10.1177/2514848620938316

International Union for the Conservation of Nature (IUCN). 2022. *The IUCN Red List of Threatened Species*. Version 2021-3. IUCN. https://www.iucnredlist.org

Jørgensen, Dolly. 2017. "Endling, The Power of the Last in an Extinction-Prone World." *Environmental Philosophy* 14: 119–138. https://doi.org/10.5840/envirophil201612542

Jørgensen, Dolly. 2018. "After None: Memorialising Animal Species Extinction through Monuments." In *Animals Count: How Population Size Matters in Animal-Human Relations*, edited by Nancy Cushing, and Jodi Frawley, 183–199. Abingdon, UK: Routledge.

Jørgensen, Dolly. 2019. *Recovering Lost Species in the Modern Age: Histories of Longing and Belonging*. Cambridge: The MIT Press.

Jørgensen, Dolly. 2021. "Erasing the Extinct: The Hunt for Caribbean Monk Seals and Museum Collection Practices." *História, Ciências, Saúde – Manguinhos* 28: 161–183.

Jørgensen, Dolly. 2022. "Portraits of Extinction: Encountering Extinction Narratives in Natural History Museums." In *Traces of the Animal Past: Methodological Challenges in Animal History*, edited by Sean Kheraj and Jennifer Bonnell, 359–375. Calgary: Calgary University Press. https://doi.org/10.1590/s0104-59702021000500007

Jørgensen, Dolly and Isla Gladstone. 2022. "The Passenger Pigeon's Past on Display for the Future." *Environmental History* 27, no. 2: 347–353.

Jørgensen, Dolly, Libby Robin, and Marie-Theres Fojuth. 2022. "Slowing Time in the Museum in a Period of Rapid Extinction." *Museum and Society* 20, no. 1: 1–11. https://doi.org/10.29311/mas.v20i1.3804

Kalshoven, Petra Tjitske. 2018. "Piecing Together the Extinct Great Auk: Techniques and Charms of Contiguity." *Environmental Humanities* 10, no. 1: 150–170. https://doi.org/10.1215/22011919-4385507

Lorimer, Jamie. 2014. "On Auks and Awkwardness." *Environmental Humanities* 4: 195–205. https://doi.org/10.1215/22011919-3614989

Manchin, Rachel. 2008. "Gender Representation in the Natural History Galleries at the Manchester Museum." *Museum and Society* 6: 54–67.

Mitchell, Audra. 2020. "Revitalizing Laws, (re)-making Treaties, Dismantling Violence: Indigenous Resurgence against 'The Sixth Mass Extinction'." *Social & Cultural Geography* 21, no. 7: 909–924.

Monson, Clark. 2005. "Cultural Constraints and Corrosive Colonization: Western Commerce in Aotearoa/New Zealand and the Extinction of the Huia." *Pacific Studies* 28: 68–93.

Naeem, Shahid, J. Emmett Duffy, and Erika Zavaleta. 2021. "The Functions of Biological Diversity in an Age of Extinction". *Science* 336: 1401–1406. https://doi.org/10.1126/science.1215855

Pálsson, Gísli. 2020. *Fuglinn sem gat ekki flogið*. Reykjavik: Bókabúð.

Poliquin, Rachel. 2008. "The Matter and Meaning of Museum Taxidermy." *Museum and Society* 6, no. 2: 123–134.

Rose, Deborah Bird, Thom van Dooren, and Matthew Chrulew. 2017. "Telling Extinction Stories." In *Extinction Studies: Stories of Time, Death, and Generations*, edited by Deborah Bird Rose, Thom van Dooren, and Matthew Chrulew, 1–18. New York: Columbia University Press.

Salmón, Enrique. 2000. "Kincentric Ecology: Indigenous Perceptions of the Human-Nature Relationship." *Ecological Applications* 10, no. 5: 1327–1332. https://doi.org/10.1890/1051-0761(2000)010[1327:KEIPOT]2.0.CO;2

Shell, Hanna Rose. 2004. "Skin Deep: Taxidermy, Embodiment, and Extinction in W. T. Hornaday's Buffalo Group." *Proceedings of the California Academy of Sciences* 55, no. 5: 88–112.

Symons, Kate and Ben Garlick. 2020. "Introduction: Tracing Geographies of Extinction." *Environmental Humanities* 12, no. 1: 288–295. https://doi.org/10.1215/22011919-8142363

Theriault, Noah and Audra Mitchell. 2020. "Extinction." In *Anthropocene Unseen: A Lexicon*, edited by Cymene Howe and Anand Pandian, 176–182. Goleta: Punctum Books.

Todd, Zoe. 2014. "Fish pluralities: Human-Animal Relations and Sites of Engagement in Paulatuuq, Arctic Canada." *Études Inuit Studies* 38, no. 1–2: 217–238. https://doi.org/10.7202/1028861ar

Van Dooren, Thom and Deborah Bord Rose. 2016. "Lively Ethography: Storying Animist Worlds." *Environmental Humanities* 8, no. 1: 77–94. https://doi.org/10.1215/22011919-3527731

Whyte, Kyle. 2021. "Indigenous Environmental Justice: Anti-Colonial Action Through Kinship." In *Environmental Justice: Key Issues*, edited by Brendan Coolsaet, 266–278. Abingdon and New York: Routledge.

Wonders, Karen. 2003. "Habitat Dioramas and the Issue of Nativeness." *Landscape Research* 28, no. 1: 89–100.

Would, Alice. 2021. "Taxidermy Time: Fleshing Out the Animals of British Taxidermy in the Long Nineteenth Century, 1820–1914." PhD thesis, University of Bristol.

2

PIGEONS AND OTHER STRANGERS IN POST-WAR BRITAIN

Philip Howell

For Kim Hui Howell

My little doves were ta'en away
From that glad nest of theirs
Across an ocean rolling gray,
And tempest-clouded airs:
My little doves, who lately knew
The sky and wave by warmth and blue.
　　　　Elizabeth Barrett Browning, *My Doves*[1]

This chapter runs the post-war history of the feral pigeon in British cities alongside that of the Black immigrants whom we now recognise as the "Windrush generation." It draws parallels between the fluctuating fortunes of these animal and human settlers, particularly insofar as enthusiastic acceptance or grudging tolerance has given way to sharper-edged suspicion of such unwelcomed "strangers." In suggesting that official policies have created more "hostile environments" for both officially unwanted pests and officially unwanted people, I argue that the space people share with birds is freighted with social and political significance. Here, the context is human and animal migration and adaptation, so that the "winged worlds" of our contemporary cities are the product of both actual and metaphorical birds of passage, but it is the historical geography of these "lines of flight" that is my theme: we are not speaking of humans and animals in the abstract, but *these* humans and *these* birds, in *these* times. As Olga Petri remarks in her introduction, the space we share with birds is always an encounter with otherness, but when that shared space is rapidly shifting, the place of human as well as animal others is put into question. Bird migration both challenges our conventional sense of *dwelling*, and when human beings as well as birds quickly come to settle permanently in another

DOI: 10.4324/9781003334767-4

place, the encounter with birds speaks reflexively to a shared search for "rootedness within uprootedness" (Clingerman 2008, 322). This chapter takes up the invitation of "winged worlds" to explore how the otherness of post-war British migrants harmonises with the otherness of pigeon "pests."

I am acutely aware of the dangers involved in considering the denigration of people and animals in the same register, and making spurious connections is probably the least of our worries. Of course, I do not want to make offensive comparisons between nonhuman animals and the racially abjected men and women whose treatment by the British state has been widely recognised as a national scandal. Nor do I want idly to gesture at the "entanglement of blackness and animality" (Bennett 2020, 5). Questions of animality and questions of race are not reducible to each other —as for instance in the alluring narrative of a common exploitation and a liberatory alliance. Instead, I want to focus on what I think of as the *switching points* between these two instances of recent cultural politics, the connections where the street pigeon and the Black migrant are brought into "fraught proximity" (Bennett 2020, 1) in the politics of post-war Britain. This *proximity* may be metaphorical or imaginative, and I want to say something about the cultural crossings of the feral pigeon and the Caribbean migrant. To this end, I turn to the postcolonial literature of the Black British migrant experience, or at least to a well-known example: the work of the novelist Sam Selvon. At other times, however, the connection between the pigeon and the migrant is direct, and as it were "live," not simply the product of imaginative intuition or tendentious hindsight. These direct connections constitute the matter of what we might think of as a "contrapuntal" animal–human history rather than ecopoetics, a shared history in which understanding the liminal status (Howell 2019) of particular human communities and that of particular animals might lead us to a more productive understanding of the present as well as the past. In other words, what can we learn from this particular encounter with human and animal otherness?

War, and peace, and the pigeon

British pigeons had a good war. The first three Dickin medals (popularly understood as the "animals' Victoria Cross") were awarded in 1943 to several pigeons serving with the Royal Air Force (RAF), and more medals have been awarded to pigeons than to all the other wartime winners combined. Domestic birds were mobilised before the outbreak of hostilities: the National Pigeon Service was founded in 1938, and the nation's pigeon breeders offered 200,000 to the war effort. Sixteen thousand five hundred and fifty-four of these birds were parachuted into occupied Europe. One of these war pigeons, a red chequer named Commando, earned his wings by flying vital information in and out of France, dodging enemy falcons and marksmen, as well as nonaligned birds of prey and the hazards of wind and weather. Commando lived to achieve honoured status in his hometown of Haywards Heath, and a blue plaque there has recently commemorated his exploits (Cusack 2016).

These were animals pressed into service, of course, and if not quite as roughly treated as in the US, where pigeons were prepared for use in an abortive missile guidance system (Skinner 1960; Capshew 1993), they can hardly be treated simply as gallant volunteers. Even so, they were mentioned in dispatches, and their war service was an established theme for many years afterwards. In 1947, for instance, the backbench MP Air-Commodore Sir Arthur Harvey must have spoken for many when he asked the Labour government to promote pigeon-keeping, on the grounds that "They played a great part in the last war, and will probably do the same in the next" (*Hansard* 25 February 1948, 447, 1921). In 1954, the pigeon's war work was still being praised, albeit with more tongue in cheek, by Hugh Fraser MP (himself a decorated intelligence officer who had been parachuted into the Ardennes):

MR. H. FRASER: I have had considerable experience of pigeons in the war. They frequently carried rude messages, but they frequently also got through when the wireless failed to do so.

MR. JACK JONES: Did the hon. Gentleman send rude ones back?

MR. FRASER: In this book to which I have referred [Osman 1950] there is a remarkable record of pigeons from the military point of view. They never once showed the white feather.

Archness aside, it is fair to say that the humble pigeon entered the second Elizabethan age with its reputation enhanced.

The end of the war promised even more, at least if we swap the pigeon for its slightly smaller but otherwise zoologically indistinguishable cousin, the dove. This branch of family *Columbidae* became the symbol of all that was hoped for in what has become known as the long peace (Chavannes-Mazel 2013), as for instance in Picasso's lithograph *La Colombe*, adopted at the 1950 World Peace Congress in Sheffield (Lewis 2014, Figure 2.1). This was surely the pigeon's finest hour, even if the Congress was Soviet-sponsored, and permanent peace frustratingly elusive.

For the other members of the family, it was all downhill from here. The dove remains ubiquitous as a *symbol*, but the humble pigeon has fluttered down from these heights of post-war popularity, or rather been knocked off its perch. In the towns and cities of post-war Britain, what has been called the public life of the street pigeon (Simms 1979) was less lofty. The city pigeon had for some time made a thorough nuisance of itself, but the descent in social status really took hold after the Second World War (Johnston & Janiga 1995, 230; Giunchi et al. 2012, 222). Pigeons moved to the cities in notable numbers, or perhaps, it is better to simply say that their flocks grew in cities after the war, even if the precise populations are uncertain (recent estimates range wildly: from 18 million in 2007 to just 550,000 breeding pairs in the UK barely a decade later).

These were not birds of high status. There is a griminess to the pigeon that makes it an icon of the run-down mundanity of post-war British life, its dour

FIGURE 2.1 Picasso, dove of peace. Detail of poster for second World Congress of the Defenders of Peace, Sheffield 1950. Author's photograph. © Succession Picasso/DACS, London 2022.

reality. It is striking that Picasso's dove (the Sheffield iteration, that is) looks distinctly scruffy, more obviously pigeon-like. This is a marked contrast both with the pure white of the real bird that Henri Matisse gifted Picasso, and with the increasingly abstract doodles, complete with olive branches, that followed. The Sheffield dove looks appropriately ragged, real, and properly urban. This is the bird with which we are familiar. In Sheffield, Picasso's sole visit to the city has recently been commemorated with life-size steel sculptures of doves by the artist Richard Bartle, gathered on the chimney stack of what is now a grill restaurant, looking down indifferently on the Peace Gardens (commemorating the appeasement of Munich 1938). These seven metal doves are from a distance indistinguishable from pigeons, and they look like the problem birds that British cities have bemoaned ever since. Sheffield City Council's own website has a familiar charge sheet: "causing damage, carrying inspect [sic] pests and leaving droppings" (Sheffield City Council n.d.). To the problems of damage, disease, and droppings, there are sometimes other problems noted: "Vocalizations of pigeons cause hysteric reactions among hypersensitive people," apparently (Haag-Wackernagel 1995, 256).

Thus, the ethereal symbol of peace has thus become no more than a nuisance and a "pest." The contentious word "feral" is one register of this unloved status. "Feral" is a wholly imprecise concept, but its indeterminacy does not stop the term endorsing a whole series of biopolitical interventions, up to and including extermination. Nicholas Holm (2020, 3) has written that "the feral has come to evoke a debased form of nature: corrupted by its time spent under human control, feral is the evil twin of wild." He rightly argues that such "ferality" is not just a matter of an in-between space, but a kind of alternation or simultaneous coexistence of "wild" and "domestic." We can see this in an early Parliamentary notice of the problem of "feral" urban pigeons. In 1953, considering the Protection of Birds Bill, and wanting to ensure that legislation on the killing of wild birds did not hamper efforts to counter the "domestic pigeon gone feral," Viscount Templewood moved an amendment that would "bring into the 'rogues' gallery' the large number of pigeons which did not seem to have any owner, and which ought to be exterminated or at least restricted." As Templewood conceded, "The rather strange wording 'gone feral' was meant to cover tame pigeons which had gone native. (Laughter)" (*Times* 11 December 1953, 3). Here, there is no pretence of precision, and it is interesting that there is a sense of the birds domesticating themselves, even whilst making themselves a nuisance, as "the inconvenient squatter of Trafalgar and Albert Squares" (*Guardian* 12 December 1953, 4), or, more recently, "the Trafalgar Square layabout" (*Daily Mail* 3 April 2014, 57). The language is instructive. "The pigeon symbolizes who we are, whereas the dove represents the 'other,' who we would like to be," writes Barbara Allen (2009, 12–13). But it is also obvious that pigeons symbolise not just ordinary life but the problems of the urban underclass. Pigeons have been associated with a whole host of "undesirables," including other "anti-social" birds (Trotter 2019). "Urban animal life, whether domestic or feral, has often been lumped in with the socially excluded – beggars, drunks and revellers and the like," argues ecologist Mike Jeffries (2018). This precipitate fall from grace has culminated in a series of high-profile campaigns against these "rats with wings" (Jerolmack 2008, 2013). Sites like London's Trafalgar Square, once famous and celebrated for its pigeons, have been made "hostile environments" for these unloved urbanites (Escobar 2014).

At other times, however, these unloved street pigeons curiously find themselves put beyond the pale in another way: by being labelled "foreign." This is not the "other" of our ideal selves, but the "them" who hone and haunt the nationalist imagination of who we think "we" are. It is a little startling for instance to see the familiar feral pigeon ranked with the Canada Goose as a disturbing "alien" (*Guardian* 20 May 1995, A. 38). One current pest control site even describes urban pigeons as "not natural to the UK" (Pest UK n.d.). It is as if these birds' canny adaptation to modern urban life, their migration from sea cliffs and country dovecotes, is enough to render them at a stroke un-British. Consider the recurrent antipathy shown towards pigeons by the late philosopher, conservative political theorist, and countryside champion Roger Scruton (*Daily Mail* 5 July 2005, 13). For Scruton, British animals are crucial to our national identity – but clearly only as long as they reside in "a pure and unsullied space of their own." The street pigeon, by contrast, is clearly in the wrong place, a verminous animal that deserves to be shot. Writing in this vein, Scruton is indistinguishable from any number of pigeon-haters in the popular press. But his commentary moves into more interesting territory when he writes:

> Unlike the songbirds that sing to us from another and more heavenly realm, or the deer that start away from us into secret arbours of their own, they look back at us with hungry eyes, standing within our world, the image of all that we have tried and failed to expel of our own lower nature.
>
> (Daily Mail *5 July 2005, 13*)

In this odd refrain, romanticism is mingled with a sense of all that has gone wrong with the world. This is a version of ferality where repellent foreignness is not very far from the surface. As a philosopher, Scruton would have remembered that for Emmanuel Levinas (1969), the face-to-face encounter with the Other is the primary ethical act, and that the defenceless eyes, in particular, plead with us: *not to kill*. Levinas suggests that animals have only a limited claim upon us – "One cannot entirely refuse the face of an animal" (see Atterton 2011, 642). But we should stress that the ethical encounter is by definition not just with those of "us" we recognise as the "same." The being whose presence we must recognise is always a "stranger," and even the "strange strangers" (Morton 2010) that nonhuman animals represent challenge us to be merciful. Probably, Scruton would have little truck with these phenomenological reflections, but he certainly had no interest in considering any responsibility towards animal others. He sees for instance in the eyes of the pigeon only illegitimate importunity, and this is itself an invitation to extermination. Scruton's attitude to street pigeons is all of a piece with an uncompromising moral hierarchy wherein those animals which "fail to elicit our sympathy" must "sink rapidly towards extinction," whilst "stranger forms of animal life" such as vermin exclude themselves as "suitors for our moral concern" (Scruton 1996, 12).

Pigeons, migrants, and the Windrush generation

What of the human "strangers" who arrived in post-war Britain? Citizens of the British Commonwealth of nations can also be said to have had a good war. Yet, when large numbers of non-white Commonwealth citizens took up their right to live, work, and settle in the home country, the response of the British state was to remap the boundaries of Britishness, tailoring citizenship to the exigencies of "managed migration" or "immigration control" (Patel 2021). Black Britons found themselves on a similarly downward trajectory, slipping down the scale from citizens to migrants, then to "immigrants," and finally (for some at least) "illegal immigrants" (see Paul 1997; Perry 2015). Too many have faced not extermination but deportation; and the spectre of "deportability" haunts far more, as it is clearly meant to in the "hostile environment" policy pioneered by the Home Office since 2009 (Wardle & Obermuller 2019).

Is there comparison between these recent campaigns against pigeons on the one hand and migrants on the other? The first thing we should note is that the urban pigeon is a relatively recent migrant too. The "magnificent, humble rock pigeon" (Allen 2009, 194), *Columba livia*, abandoned its natural habitats generations ago, domesticated ancestors escaping into urban environments where buildings provide perfect substitutes for sea cliffs. The supply of food from human society was also perfectly ample for their relatively small populations, and they became such familiar strangers that their strangeness is sometimes hard to register. Notwithstanding what I have written above, "feral" pigeons have become so established that it takes a certain effort of will to represent them as non-native newcomers. Banksy's 2014 Clacton-on-Sea mural depicting a drab handful of pigeons instructing a colourful parakeet to "go back to Africa" and to "keep off our worms" (Figure 2.2) is a mille-feuille of irony. The accusations of racism this mural prompted (see Johnston 2014) are particularly risible and indicate in themselves the poverty of the public debate on immigration and its history in Britain.

The less dim-witted will recognise that we need to resist the idea that Britain is a "nation-state" whose "indigenous" subjects have only latterly been supplemented by immigrant strangers. Engin Isin's (2015) alternative term "empire-state" is much more useful when thinking about the history of migration to and from the British Isles, and whilst the pigeons hardly cared whether they arrived in a nation-state or an empire-state, it is vital to understand that Britain was an imperial metropole, and London an imperial metropolis, long before the pigeons multiplied. Trafalgar Square was the heart of this heart of Empire – its "hub," as the *Illustrated London News* called it in 1927, using a painting by Charles Turner (Figure 2.3). The presence of pigeons in Trafalgar Square makes them a kind of imperial subject, even

FIGURE 2.2 Banksy mural, Tendring District Council boathouse, Clacton-on-Sea, c.2014. Reproduced with permission.

FIGURE 2.3   Charles E. Turner, "The hub of the British Empire: feeding the pigeons
in Trafalgar Square," *Illustrated London News*, 12 March 1927, 442–443.
© Illustrated London News/Mary Evans.

if this attribution seems far-fetched because of their discountable ordinariness. In
Turner's picture, feeding the pigeons in the Square is a pleasant pastime, the children if not the adults oblivious to the pomp and grandeur of this iconic imperial
site. There is both comfort and irony in the fact that this domestic scene is played
out in a setting that attests to Britain's great imperial reach (Mace 1976). London's
pigeons play their part in this theatre of empire.

In Turner's painting, there is no hint of the horrors to come, nor the imminent
influx of visitors and strangers and the permanent post war settlement of hundreds
of thousands of migrants. Nor are the capital's pigeons portrayed as "pests." The
homely theme is continued after the war, and the reassurance of normality and
peaceful coexistence is surely the reason why a generation of artists returned to
the business of feeding the pigeons in the heart of London. There are many examples, but I might single out Harold Blackburn's 1947 work, *Trafalgar Square*, where
Christ walks amongst the pleasure-seekers, and for all his exotic dress (never mind
the halo), attracting less attention than the plentiful pigeons. In Blackburn's picture, Landseer's imperial lion is equally domesticated, and here, the lion seems to
lie down with the lamb, and pigeons and citizens live and flourish alongside each
other, and all is well with the world. Blackburn's painting predates the arrival of the
Empire Windrush by a single year, and London's citizens are a pretty monochrome
bunch. But another artist, Harold Dearden, does register the changing realities of
post-war London, in a pair of paintings portraying non-white visitors or citizens
engaged in the same iconic pursuit (Figure 2.4). Whilst the Trafalgar Square theme
is much the same, this picture has a new racial narrative: this is a domestic idyll in
the heart of London, but it is not just birds and humans living in harmony, but

FIGURE 2.4 Harold Dearden, Caribbean Family in Trafalgar Square. Oil on canvas, 1950–1962. Museum of London, 99.115/1. Copyright holder unidentified.

black and white too. I might be overreading, but the pigeon at the bottom of the frame has its head enclosed by the diamond pattern of the central figure's dress, in what looks to me like the orthodox icon of the Holy Spirit. Here, again, the pigeon stands in for the dove, its grey and grimy mundanity bleached out by the Christian symbolism.

Of course, this is heavily romanticised. I do not want to concentrate here on the racism directed at the Windrush generation, and all the non-white migrants who followed them, appalling as it is (see Olusoga 2016; Paul 1997; Perry 2015), but rather to look at the relations between human beings and birds at this particular conjuncture, when the empire comes home to roost.

Consider an episode from Sam Selvon's 1956 novel *The Lonely Londoners* (Selvon 2006). Selvon's postcolonial perspective on the migrant experience in post-war London is well-known, but what is less appreciated are his asides on the experience of migrant men and women living amongst their animal neighbours as well as the variously curious or hostile white ones. In *The Lonely Londoners*, the young Trinidadian "Sir Galahad" (real name Henry Oliver) is newly arrived in London. He is sunnily optimistic but immediately challenged by the hard winters of London and so straitened by hunger that he conceives a plan to catch a pigeon from the local park for his supper. Galahad does not think of the pigeons as fellow migrants, but he does still wonder where they come from, and he notes, with some chagrin, that the animals in Britain, even the dim and dusky pigeons, are treated better than people like him:

> It does have a lot of them flying about, and the people does feed them with bits of bread. Sometimes they get so much bread that they pick and choosing, and Galahad watching them with envy. In this country, people prefer to see man starve than a cat or dog want something to eat.
> Watching these fat pigeons strut about the park, the idea come to Galahad to snatch one and take it home and roast it.
>
> *(Selvon 2006, 117)*

This cunning plan, so unremarkable in Port of Spain and so nefarious in London, has to be carried out with careful ornithological observation, and a military precision bordering on espionage: "he wasn't so much frighten of the idea of the snatch as what would happen if one of them animal-loving people see him" (Selvon 2006, 118). Predictably, Galahad attracts unwonted attention:

> Galahad eye a fat fellar who edging up to the rail. He start to drop bread a little nearer, until the bird was close. He make the snatch so quietly that the other pigeons only flutter around a little and went on eating. He start to swing the pigeon around, holding it by the head, for he want to kill it quick and push it in his pocket.
> As aforesaid, that particular season it was as if the gods against the boys, and just as Galahad swinging the pigeon one of them old geezers who does always wear furcoat come through the entrance with little Flossie on a lead, to give the little dear a morning constitution, and as soon as Flossie spot the spade she start a sharp barking.
> 'Oh you cruel, cruel beast!' the woman say, and Galahad head fly back from where he kneeling on the ground to handle the situation better. 'You cruel monster! You killer!'
> Galahad blood run cold: he see the gallows before him right away and he push the pigeon in his jacket pocket and stand up, and the pigeon still fluttering in the pocket.
> 'I must find a policeman!' the woman screech, throwing her hands up in the air, and she turn back to the road.
>
> *(Selvon 2006, 118–119)*

The resulting meal of pigeon and rice is bought at the expense of Galahad's guilty conscience, but it also points out the distance between the "winged worlds" of his native Caribbean and his adopted Britain. Pigeons may have been wearing out their welcome for the authorities, fed up with the clean-up bills, but they were familiar urban residents, and largely accepted by the general public, and even acknowledged as worthy of protection. Galahad's friend Moses reminds him of the risks he has run, in a country where animal welfare has become a kind of shibboleth:

> Now when Galahad did reach back home, and he sit down and start to pick the bird feather, he start to feel guilty. All he try to argue with himself that he only do it because he hungry and things brown, still the feeling that he do a bad thing wouldn't leave him. 'What the hell I care,' he say to himself, 'so much damn pigeon all about the place. Look how they making mess all about in Trafalgar Square until the government trying to get rid of them. What the hell happen if I snatch one to eat?' But when he finish plucking the pigeon conscious humbugging him so much that he fling it in a corner. Little later he thought about Moses. 'Moses in this country long,' he say to himself, 'and if he could eat it I don't see why I must feel so guilty.' So he went round by Moses.
> 'I make a snatch,' Galahad say, and he tell Moses all what happen.
> 'Boy, you take a big chance,' Moses say. 'You think this is Trinidad? Them pigeons there to beautify the park, not to eat. The people over here will kill you if you touch a fly.' But all the same the old Moses eyeing the bird, and is a nice, fat one.
>
> *(Selvon 2006, 119–120)*

Selvon mines this business for all its comic worth, and there is obviously a pointed contrast between the mores of Trinidad and those of Britain. But the point is not that British people – white British people, that is – are stupidly sentimental, and Caribbean people – Trinidadians in this case – are straightforwardly sensible in how animals should be treated and how they should be cooked. Instead, we are guided to an incipiently modern, multicultural London where human–animal relations are the source of class and racial friction (see Kim 2015). Migration brings people together, but it does not automatically produce a pleasing picture of human/human and human/animal harmony. Ecologists speak of "commensality" – literally, animals and humans eating from the same table – but instead of "commensal" pigeons being happily fed by human hosts we have the pigeon ending up as food for the hungry migrant, to the outrage of the "animal-loving" Brit. What matters here is how the relations between people and animals reflect the potential tensions between different human communities. Galahad's pigeon and rice is more than just a meal.

The war against commensality

Selvon's episode is fiction, but that it is not *just* fiction is confirmed by a newspaper story from 1953 (*Times* 8 March, 1957, 12): "Man killed pigeons in Trafalgar Square: £5 fine for cruelty." A very few details followed: Eric McKenzie, aged 47, stoker, native of British Guiana, was fined for causing unnecessary suffering to two pigeons in Trafalgar Square. An Inspector W. Grant reflected: "I think he must have fed them and then seized them. McKenzie said he was out of work and hungry." In its own way, this sparse information offers a poignant counterpart to Selvon's comic brio. But it underlines the fact that post-war migrants' eating habits (see for instance Wills 2017, 236) were a signal part of their "strangeness" to the resident white British. Who eats what, how, and when remains a central issue for postcolonial cultural politics, and food divides us as much as it brings us together (see Cook and Harrison 2003). More broadly, the ideal of "commensality" is always conditional, bounded, and exclusive (Kerner, Chou & Warmind 2015). If we look further at both the eating and the feeding of pigeons, the cultural politics of both human and more-than-human commensality in post-war Britain is more clearly revealed, as are the limits to the hospitality encoded in the ideal of eating together. I note here that pigeons quickly fall off the menu after the war, and feeding them too would also become an offence, as part of the "pigeon wars" of the 1990s and after. Feeding pigeons and feeding on pigeons tested the ideal of commensality, community, and society, since sharing food implies the idea of sharing the table, sharing *place* (Jönsson, Michaud and Neuman 2021). Instead of offering food to strangers, we have what has been called "segregative commensality" (Grignon 2001), where eating together serves to draw lines between "us" and the "strangers" among us.

What is obvious from Eric McKenzie's offence, for instance, is that Trafalgar Square's pigeons were by 1957 off the menu. Killing pigeons for food fell foul of the established animal welfare injunction against causing "unnecessary suffering." This, despite the fact that pigeons have long had both an edible and an ornamental past. The wild woodpigeon, along with the rook, was fair game, particularly during the war, especially since it was blacklisted for feeding on scarce grain. Stomachs may even have steeled themselves to the street pigeon, when necessity came calling. A central character in Michael Moorcock's *Mother London* (1988) recalls his grandmother living in the war on pigeons caught for her in the street by her cat; but given that he is an outpatient from a mental hospital as well as a fictional being, we might not want to take this for literal truth. Roger Scruton (*Daily Mail* 5 July 2005, 11) is unequivocal however: "When food was short during the war, feral pigeons in London ended up in that most delicious of traditional English dishes, pigeon pie." It is just possible that Scruton might have been thinking of an old episode of the long-running WWII homefront sitcom *Dad's Army* (1968–1977), where the enterprising spiv Private Walker supplements Corporal Jones's butcher's shop with unorthodox supplies, the sudden abundance of fresh poultry coinciding with a reported decline in Trafalgar Square's pigeon population. But whether there was much pigeon predation during the war is not important. Much more significant is the taking of pigeons, in peacetime, from a place where they were popularly associated and largely accepted: this was McKenzie's offence. As a recent migrant, unconscious of the affection that most people seem to have nurtured for the urban pigeon, he might have been forgiven for confusion on this score.

In Selvon's radio dramatisation of this pigeon business, the contrast between the "plump, importunate pigeons" (*Times* 25 October 1969, 3) of Trafalgar Square and the hungry Black migrant is given even more emphasis. The character Cap, who tells everyone that he is a Nigerian prince, having listened to Galahad's pigeon escapade, is maddened by the fact that pigeons are provided handouts:

> CAP: They only flying about the place and not serving any useful purpose. They congregate near the Bayswater Road, where them old geezers feed them with bread and biscuits.
>
> *(Selvon 2008, 90)*

It is not only the ready opportunities to eat pigeons that are significant, then; it is also the fact that pigeons are *fed* when others go hungry. There seems to be, in this commensality, a form of social communion from which Cap and Galahad and others are definitively excluded. In Britain, pigeons are fattened, not eaten: that seems to be part of what defines the Britishness from which these migrants are shut out. What is striking in the subsequent decades is the reinforcement of the message that eating (street) pigeons is wrong, but also the increasingly peremptory condemnation of the practice of feeding them. The "hostile environment" long confronting Black Britons finds another parallel in the attempt to thin the numbers of pigeons, not least by policing the habit of feeding them.

As we have suggested, the post-war British public were relatively relaxed about these avian neighbours. Pigeon experts in the 1950s and 1960s concluded that "the generally harmless or friendly attitude of mankind towards them does ... enable London's pigeons to exploit all possible feeding areas" (Goodwin 1960, 202), noting that municipal projects of pigeon extermination were likely to be frustrated by "the casual good nature of ordinary people" (Gompertz 1957, 12). As late as 1979, Eric Simms could write that "the poor, the old and the young may receive a real uplift in spirit from feeding the street pigeons or just watching them going about their daily lives" (Simms 1979, 8). The geographer Shawn Bodden (2021) has rightly remarked on the relative rarity of "pigeon problems," for all the cultural denigration: he points profitably to the *negotiations* of shared urban space in which pigeons play a part and emphasises the fact that pigeons and humans largely get along, even if "pigeon-proofers" and other antagonists aim to deter the birds' presence. I would only add that these negotiations are coloured by racial and social antagonism.

It is the war against the pigeon that is significant in recent years, and we might think of this as a war *against commensality*, given that feeding pigeons – that once-innocent gesture – has become identified as the problem at hand. Municipal authorities have for decades directed their ire against the ranks of well-intentioned pigeon feeders. A BBC news report from 1999 cited the Spokesman of the Tidy Britain Group:

> Research seems to suggest that the best way of dealing with it is to take away the pigeons' food source. We hope that people will think before they feed pigeons. It is not natural for pigeons to rely on human beings for their food, and we now have populations of lazy urban birds.
>
> *BBC News 1999*

Sometimes, it is city-dwellers generally who are blamed, for their crowded numbers and wasteful food habits; for Roger Scruton, the pigeon is (inevitably) just another example of the "heaving throng of parasites" that feast off the leavings of a throw-away society (*Daily Mail* 5 July 2005, 13). Feral pigeons do tend to concentrate in "circumscribed areas where much bread and other artificial food (varying from cheese and chocolate to peas and potato chips) is provided" (Goodwin 1960, 205). But as often as not, it is the deliberate feeding of pigeons that has become offensive to the authorities. Quoting a town centre official, the 1999 BBC report also blamed the soft-heartedness of the pigeon feeder as a kind of emotional or mental aberration: "Effectively, pigeons are airborne rats, but you always get one sad person who goes into town every day to feed the pigeons, and the best policy is to ignore them" (BBC News 1999). The gesture of feeding the pigeons, which seemed not only innocent but also admirable in the paintings discussed previously, has become a kind of psychological or social pathology.

This is what Mayor of London Ken Livingstone took aim at when he revoked the licence of the last pigeon food seller in Trafalgar Square, Bernard Rayner. The days of feeding the pigeons in Trafalgar Square were numbered. For the first time, British authorities had succeeded in establishing the point that *feeding* pigeons was wrong. But this "war against the pigeon" was accompanied by dire warnings about *eating* pigeons too. Before the 1990s, pigeons were off the menu because the public simply wouldn't stand for it. In the aftermath of the "war against the pigeon," the press reported the surreptitious thinning of London's pigeon population by unscrupulous criminals feeding unsuspecting restaurant patrons on illicit, and dangerous, flesh. Whilst the fictional Galahad and the real Eric McKenzie were guilty only of giving way to their appetites, these pigeon nappers were endangering others. In the 1990s, Londoners would be warned off gastronomic pigeon on grounds of public health, their flesh "so polluted that it was poisonous" (*Times* 18 November 1997, 83). This advice was doled out in response to the possibility that restaurants were unscrupulously sourcing the "scrawny London feral pigeon" rather than the "plump Norfolk woodpigeon" (*Times* 11 March 1996, 7). A mysterious figure, the "phantom birdnapper of Trafalgar Square," was supposed to have taken 4000 pigeons to sell on, sometimes nabbing them in broad daylight (*Times* 9 March 1996, 21). As the *Times* editorialised:

> For pigeons to be fed on tourist-trap grain and Soho black bin bags, and then to be recycled through the Soho restaurants might seem a virtuous cycle. But urban pigeons can be unhealthy immigrants, and infect their eater with diarrhoea or even terminal belly-ache. Demand from restaurants for suspiciously cheap pigeons is not likely to last.
>
> (Times *9 March 1996, 21*)

These pigeons were not only "unhealthy immigrants." A shadow of suspicion fell upon other strangers. One suspected pigeon-napper claimed to be collecting the birds for a Peckham pigeon racing club, though local beat officer Roy Riggs claimed that "there is a strong suspicion these pigeons are turning up in pies rather than in races. They are probably being sold on to Greek pie-making restaurants as they are some sort of delicacy out there" (Carroll, 1996, 7). The "pigeon-napper" turned out to be a Jason Lidbury, unemployed painter, paid 20p per bird by his uncle, though he was neither sure nor curious about what would happen to them: "You can't do anything with them, they're just vermin" (Carroll, 1996, 26). The police were alerted by long-time pigeon-feed seller Bernard Rayner, who claimed that Lidbury was damaging his trade. Rayner flew the flag for the harmless practice of feeding the pigeon, whilst denouncing the surreptitious practice of feeding off them. However, some better-connected pigeon-haters humorously applauded Lidbury's enterprise:

> Baroness Trumpington: My Lords, although I am entirely in favour of private enterprise, does the noble Lord agree that a health risk could be posed from the practice carried out by some who trap the pigeons with nets and then flog them to restaurants? I do not believe that eating Trafalgar Square pigeons can be good for one's health, or at least not for mine.
>
> (Hansard, *12 June 2000, 613, 1371*)

Conclusions

I have regularly invoked the term "hostile environment." At the risk of facile comparisons, I have suggested a parallel concern over unwanted "strangers" in post-war Britain: the rise of the street or feral pigeon as an urban pest, and the Windrush generation and their successors who in their thousands brought the Black Commonwealth home to the imperial metropole, only to be treated as more or less unwelcome "immigrants." The link between feral animals and marginalised human communities has been well attested (Fine and Christoforides 1991), and Maan Barua (2021, 12) has recently written that "London has witnessed a surge in hostile architecture targeting the unhoused and the homeless, and its extension to include feral birds, from pigeons to parakeets, is striking." The "often rejected, sometimes reviled, street pigeon" (Simms 1979, 9) is a ready emblem of the "feral" underclass: "greedy, dirty and streetwise," "the shabby grey unemployables of city life" (*Times* 18 November 1997, 83). The suggestion that there are links between British society's response to the urban pigeon and the Caribbean migrant may strike the reader as more fanciful, but I hope to have shown that in the "eco-poetics" of the Black diaspora there are points of contact as well as comparison. Specifically, I have shown how the novelist Sam Selvon articulates the precarity of the Black migrant through the presence of the urban pigeon. The winged world of the street pigeon is brought fleetingly into contact with our social and cultural politics of migration and the racism on which it was built as well as that which it provoked.

Roger Scruton, the arch-conservative and enthusiastic Brexiteer, who Jonathan Portes has argued provided contemporary Toryism with a "licence for bigotry" (Portes 2020), gives us a sense of this complex cultural politics, even as he reflects dyspeptically on the nature of migration and its inversion of dwelling. In his novella *A Dove Descending*, Scruton confronts his heroine Zoë, daughter of the immigrant Cypriot Yannakis Costas, with both idealised birds and the debased reality of the feral migrant. For Zoë, Yannakis's pigeons, including his favourite Evyenia, are "symbols of the outer world and its variety, messengers sent across the roofs and fields of England, preparing the day when they too would take wing." In their wish to leave for the sunny skies of the Aegean, these pigeons seem to agree with the conventionally radical Leacock, Zoë's tutor and admirer, who admires everything "alien" and despises everything "England." But at one point, Zoë is surrounded by the importunate beggars that pigeons have become in the metropolis:

> A surging blue mass of verminous pigeons had gathered around her, clawing one another, outlining Zoë's legs as though to conjure into the space they occupied some more beneficent existence. She looked down at them with loathing, recalling the runculations of the dovecote, and feeling disbelief that a single species could produce Evyenia and these. The cold beads of their eyes reflected nothing – no love or interest had found itself in them, no dream had shone back from their depths.
>
> *(Scruton 1991)*

In Scruton's inveterate hatred of the street pigeon, the eyes of these debased birds are wells of darkness, with no trace of the "glittering eyes" that for Elizabeth Barrett "showed their right/To general Nature's deep delight."[2] Scruton seems here to despise what migration does, to birds – and to people – and he certainly prefers the unpolluted purity of rural England or an idealised magna Graecia.

Street pigeons are here to stay, however. Like Black Britons and other migrants, like my mother, they are "an integral part of urban life" (Simms 1979, 85). Like all migrants, they provide "another model of how to dwell in place" (Clingerman 2008, 315). Neither are strangers anymore, though they continue to be met with policies that treat them as nothing more. In these precarious times, perhaps we can remember Elizabeth Barrett's orison, directed to her two beloved doves, transported from the Caribbean to a grey urban captivity on the other side of a pitiless ocean:

> So teach ye me the wisest part,
> My little doves! to move
> Along the city-ways with heart
> Assured by holy love,
> And vocal with such songs as own
> A fountain to the world unknown.
>> Elizabeth Barrett Browning, *My Doves*[3]

Notes

1 See Barrett Browning (2021), but note that the orthography differs from edition to edition. The poem was written about two turtle doves given to Elizabeth Barrett by her father and published in 1838. For the complex connections between these birds' Jamaican origin and the Barrett family's connection with transatlantic slavery, see Parker (2020).
2 From Barrett Browning (2021), "My doves."
3 The "fountain to the world unknown" takes us to the *infinity* that allows us to acknowledge and accept the claims of strangers and the strange world they inhabit (Levinas 1969).

References

Allen, B. 2009. *Pigeon*. London: Reaktion.

Atterton, P. 2011. Levinas and our moral responsibility toward other animals. *Inquiry*, *54*(6): 633–649. DOI: 10.1080/0020174X.2011.628186.

Barrett Browning, E. 2021. *The swan's nest among the reeds: Selected bird poems of Elizabeth Barrett Browning*. Bristol: Ragged Hand.

Barua, M. 2021. Feral ecologies: The making of postcolonial nature in London. *Journal of the Royal Anthropological Institute*, 1–24. DOI: 10.1111/1467-9655.13653.

BBC News. 1999. UK pigeons, not a problem to poo-poo. *BBC News*, 19 January 1999, http://news.bbc.co.uk/1/hi/uk/257284.stm

Bennett, J. 2020. *Being property once myself: Blackness and the end of man*. Cambridge, MA: Harvard University Press.

Bodden, S. 5 January 2021. Pigeon problems: Sharing space in the multispecies city. *Winged Geographies* blog, https://www.wingedgeographies.co.uk/post/pigeon-problems-sharing-space-in-the-multispecies-city

Capshew, J. H. 1993. Engineering behavior: Project pigeon, World War II, and the conditioning of B. F. Skinner. *Technology and Culture*, *34*(4): 835–857. DOI: 10.2307/3106417.

Carroll, H. 1996. Pie from the sky as thief strikes in Trafalgar Square. *Daily Mail* 8 March 1996, 7.

Chavannes-Mazel, C. A. 2013. Dove: Visual arts. In Klauck, H-J., Leppin, V., McGinn, B., Seow, C-L., Spieckermann, H., Walfish, B. D., & Ziokowski, E. J. (Eds.), *Encyclopedia of the Bible and its reception, Vol. 6: Dabbesheth - dreams and dream interpretation*, 1127–1129. Amsterdam: De Gruyter.

Clingerman, F. 2008. The intimate distance of herons: Theological travels through nature, place, and migration. *Ethics Place and Environment*, *11*(3): 313–325. DOI: 10.1080/13668790802559718.

Cook, I., & Harrison, M. 2003. Cross over food: Re-materializing postcolonial geographies. *Transactions of the Institute of British Geographers*, *28*: 296–317. DOI: 10.1111/1475-5661.00094

Cusack, B. 2016. Second world war hero Commando gets a blue plaque – and he's a pigeon. *Daily Mirror*, 23 March 2016, https://www.mirror.co.uk/news/uk-news/second-world-war-hero-commando-5387701

Escobar, M. P. 2014. The power of (dis)placement: Pigeons and urban regeneration in Trafalgar Square. *Cultural Geographies*, *21*(3): 363–387. DOI: 10.1177/1474474013500223.

Fine, G. A., & Christoforides, L. 1991. Dirty birds, filthy immigrants, and the English Sparrow war: Metaphorical linkage in constructing social problems. *Symbolic Interaction*, *14*(4): 375–393. DOI: 10.1525/si.1991.14.4.375.

Giunchi, D., Albores-Barajas, Y. V., Baldaccini, N. E., Vanni, L., & Soldatini, C. 2012. Feral pigeons: Problems, dynamics and control methods. In Soloneski, S., & Larramendy, M. L. (Eds.), *Integrated pest management and pest control: Current and future tactics*, 215–239. Rijeka: InTech.

Gompertz, T. 1957. Some observations on the feral pigeon in London. *Bird Study*, 4(1): 2–13. 10.1080/00063655709475863.

Goodwin, D. 1960. Comparative ecology of pigeons in inner London. *British Birds, 53*(5): 201–212.

Grignon, C. 2001. Commensality and social morphology: An essay of typology. In Scholliers, P. (Ed.), *Food, drink and identity: Cooking, eating and drinking in Europe since the Middle Ages*, 23–33. Oxford: Berg.

Haag-Wackernagel, D. 1995. Regulation of the street pigeon in Basel. *Wildlife Society Bulletin*, 23(2): 256–260. DOI: 10.2307/3782800.

Holm, N. 2020. Consider the (feral) cat: Ferality, biopower, and the ethics of predation. *Society & Animals* (online publication date 27 April). DOI: 10.1163/15685306-BJA100006.

Howell, P. 2019. The trouble with liminanimals. *Parallax, 25*(4): 395–411. DOI: 10.1080/13554645.2020.1731007.

Isin, E. F. 2015. Citizenship's empire. In Isin, E. F. (Ed.), *Citizenship after orientalism: Transforming political theory*, 263–281. London: Palgrave.

Jeffries, M. 2018. We've saved pink pigeons from extinction - now let's be kinder to their grey cousins. *The Conversation* 4 December 2018, https://theconversation.com/weve-saved-pink-pigeons-from-extinction-now-lets-be-kinder-to-their-grey-cousins-107849

Jerolmack, C. 2008. How pigeons became rats: The cultural-spatial logic of problem animals. *Social problems, 55*(1): 72–94. DOI: 10.1525/sp.2008.55.1.72.

Jerolmack, C. 2013. *The global pigeon*. Chicago: Chicago University Press.

Johnston, C. 2014. Council removes Banksy artwork after complains of racism. *Guardian*, 1 October.

Johnston, R. F., & Janiga, M. 1995. *Feral pigeons*. Oxford: Oxford University Press.

Jönsson, H., Michaud, M., Neuman, N. 2021. What is commensality? A critical discussion of an expanding research field. *International Journal of Environmental Research and Public Health, 18*(12): 6235. DOI: 10.3390/ijerph18126235.

Kerner, S., Chou, C., & Warmind, M. 2015. *Commensality: From everyday food to feast*. London: Bloomsbury.

Kim, C. J. 2015. *Dangerous crossings: Race, species, and nature in a multicultural age*. Cambridge: Cambridge University Press.

Levinas, E. 1969. *Totality and infinity: An essay on exteriority*. Pittsburgh: Duquesne University Press.

Lewis, R. 2014. The dove: Picasso and Matisse. *Lewis Art Café blog*, March, http://lewisartcafe.com/the-dove-picasso-and-matisse/

Mace, R. 1976. *Trafalgar Square: Emblem of empire*. London: Lawrence & Wishart.

Moorcock, M. 1988. *Mother London*. London: Secker & Warburg.

Morton, T. 2010. Thinking ecology: The mesh, the strange stranger, and the beautiful soul. *Collapse, VI*: 265–293.

Olusoga, D. 2016. *Black and British: A forgotten history*. London: Macmillan.

Osman, W. H. 1950. *Pigeons in World War II*. London: Racing Pigeon Publishing Company.

Parker, E. T. 2020. Birds in literature: Barbary Dove. *A Finer Grace blog*, June, https://www.afinergrace.com/post/birds-in-literature-barbary-dove

Patel, I. S. 2021. *We're here because you were there: Immigration and the end of empire*. London: Verso.

Paul, K. 1997. *Whitewashing Britain: Race and citizenship in the post-war era.* Ithaca: Cornell University Press.

Perry, K. H. 2015. *London is the place for me: Black Britons, citizenship, and the politics of race.* Oxford: Oxford University Press.

Pest UK (n.d.). Pigeons, pests? https://www.pestuk.com/blog/pigeons-pests/

Portes, J. 2020. Roger Scruton's brand of conservatism became a licence for bigotry, *Guardian* 17 January 2020. https://www.theguardian.com/commentisfree/2020/jan/17/roger-scruton-conservatism-bigotry-anti-immigrant

Scruton, R. 1991. *A dove descending.* London: Sinclair-Stevenson.

Scruton, R. 1996. *Animal rights and wrongs.* London: Demos.

Selvon, S. 2006. *The lonely Londoners.* London: Penguin.

Selvon, S. 2008. *Eldorado West One.* Leeds: Peepal Tree Press.

Sheffield City Council. (n.d.). Pest control advice and information. https://www.sheffield.gov.uk/home/public-health/pest-control-advice

Simms, E. 1979. *The public life of the street pigeon.* London: Hutchinson.

Skinner, B. F. 1960. Pigeons in a pelican. *American Psychologist, 15*: 28–37. DOI: 10.1037/h0045345.

Trotter, S. 2019. Birds behaving badly: The regulation of seagulls and the construction of public space. *Journal of Law and Society, 46*(1): 1–28. DOI: 10.1111/jols.12140.

Wardle, H., & Obermuller, L. 2019. "Windrush generation" and "hostile environment": Symbols and lived experience in Caribbean migration to the UK. *Migration and Society, 2*(1): 81–89. DOI: 10.3167.arms.2019.020108.

Wills, C. 2017. *Lovers and strangers: An immigrant history of post-war Britain.* London: Penguin.

3

MIGRATION AT THE LIMIT

More-than-human creativity and catastrophe

Andrew J. Whitehouse

Introduction

Migration for birds is both an ordeal and a response to ordeal. Shifting ecological conditions make these ordeals more profound, particularly for birds undertaking remarkable journeys to escape uninhabitable Arctic winters. New developments in monitoring and tracking birds have enabled researchers to learn more about their migrations and changing strategies. Although scientists often explain migration as arising primarily from various genetically influenced orientations that compel birds to respond to environmental conditions such as light and magnetic fields in specific ways (Newton 2008; Lees and Gilroy 2022), my interest here is in how birds experience migration and in the potential for them to respond to its ordeals creatively.

To do this, I use the concept of 'the limit', which Paolo Maccagno (2019) has developed to understand the creative potential of profound bodily ordeals such as running marathons. Birds migrating, just like marathon runners, are moving at the edge of physical and ecological limits and can hit a metaphorical wall at a certain point. The metaphor of the wall describes how runners can respond to this experience of their body at its limit. They can attempt to run through the wall, which quickly leads to bodily collapse, or they can run alongside it, running not with a goal in mind but, as Maccagno says, as if running forever.

According to Maccagno:

> The limit, rather than being a border or a separating line, is a space with high educational potential, which allows one to become exposed and to 'cut through the world'. [This] notion can overcome the idea of a linear progression between statuses, and highlight a movement in-between, foregrounding presence rather than identity.

(2019, v)

DOI: 10.4324/9781003334767-5

The 'in-between-ness' of migration and its potential limits render it a creative process that birds never entirely leave. For migratory species, life is lived along long, ongoing lines of flight that sometimes lead across continents and sometimes gather, during breeding and wintering for example, in more settled bundles of daily movements.

This notion of the limit, I argue, provides a means of reframing bird migration. Migration can readily be thought of as the responsive movement of birds within the limits of where life is possible for them. These limits, such as seasonal changes that make inhabitable places uninhabitable, are thus influencing their movements. These movements in themselves, however, generate further limits, since birds will normally have to move through uninhabitable areas to get to where they can survive. For example, they regularly need to traverse seas or deserts that place their lives under constant threat. These remarks appear straightforward, but they change the emphasis in how migration is understood in the three important ways described below.

First, the limit is emergent from the indivisible relationship of organism and environment. Whatever the limits of life are for an organism are contingent on its bodily capacities to inhabit its ever-changing environment. The organism and environment are not understood as being additive, in the sense of limited environmental conditions created in advance and a genetically determined body with its own specified capacities (cf. Ingold 2000, 22). Rather, those limits are not 'in the environment' but are revealed in relation to the life and movements of the bird.

Second, the emphasis the limit places on responsive movement counters the conventional assumption of migration as a pre-determined movement between two places through navigation. Instead, migration is a form of wayfaring in-between (Ingold 2000, 273–304; 2007, 72–103). Migration routes are thus replaced by lines of flight that emerge in relation with the limits of life. It is, of course, entirely normal for conventional approaches to bird migration to plot the migration as a line on a map. But for Ingold, the type of line involved is central to how movement is understood. For him, there is a pivotal contrast between point-to-point lines of transport, in which the line exists between two pre-existing nodes like a line in a dot-to-dot drawing or stations in a railway network, and the emerging lines of movement in life that are traced by beings as they carry on in the world. If the lines of migration shift from the former to the latter, then ways of thinking about migration can change with it.

Third, the 'in-between' quality of migration establishes it as both a dangerous and creative process at the edges of inhabitation. Ever-shifting limits can destroy birds in catastrophic conditions that are uninhabitable. They may also lead to new relations with limits that encourage new ways of inhabiting the world and new lines of flight. Rather than birds following pre-determined routes, migration is an opening up of possible paths. Going beyond the limits of endurance and knowledge is likely to generate new paths.

In this chapter, I examine the potential of the limit to understand migration in new ways. The aim is, to paraphrase Vinciane Despret (2016), to consider what birds might say about migration if we asked the right questions. I do this through a series of case studies that connect my own encounters with bird migration and its limits with scientific research on the same birds. These encounters with limits begin with migratory geese in Scotland and Japan, before continuing to Yellow-browed Warblers, which appear in ever greater numbers in northwest Europe in autumn. Finally, I consider the case of the endangered Spoon-billed Sandpiper, a rare and distinctive wading bird that appears on the brink of catastrophe in east Asia.

In exploring these cases, I take inspiration from the ecological philosopher Val Plumwood. Plumwood famously survived an attack by a crocodile, and this experience forged much of her later thinking. She writes:

> Some events can completely change your life and your work… They can lead you to see the world in a completely different way, and you can never again see it as you did before. You have been to the limit, and seen the stars change their course. That extreme heightening of consciousness evoked at the point of death is… of a most revelatory and life-changing kind—for those who, against all odds, are given a reprieve and survive.
>
> *(2012, 11)*

Plumwood connects the limit to death and the catastrophic. Prior to its modern usage to mean a sudden and emphatic disaster, 'catastrophe' was used to describe the conclusion, often tragic, of a drama. The Greek origins of the word, however, connote an overturning. Catastrophes can thus be an ending or a new beginning and perhaps can be both. The limit certainly presents both possibilities, as is apparent from Plumwood's example. The limits of life and the possibility of death become perceptible in catastrophes but can generate new possibilities for living. The parallels with bird migration are immediately apparent. Migration could arise *from* and give rise *to* the creative possibilities of the limit. How can this help rethink migration? And what limits that birds experience are beyond their capacity of response? When do the ordeals of life become catastrophic and for whom?

Limit 1: Tracking movements

4 October 2020. Aberdeen, Scotland

The day begins full of promise, as easterly winds and overnight rain are perfect for bringing migratory birds along the coast. I soon find small birds sheltering in the thin patches of cover, but more conspicuous are noisy skeins of geese passing southwards. Barnacle Geese are on the move, yapping constantly as they pass overhead. Presumably, these are birds heading to the Solway Firth in southern Scotland. They will have travelled over the North Sea from Norway, their line of flight moving

FIGURE 3.1 Barnacle Geese over Girdle Ness, Aberdeen 4 October 2020.

from breeding grounds in Svalbard. I have never seen such numbers here, and I count 630 by the end of the day.

Like all geese, Barnacles are intensely social. A solitary goose is an unhappy one. Migration is done in large congregations, often involving family parties. Geese are also vocal as they fly; even at night time, their calls trace a path of sound through the air, perhaps drawing others into their throng. Their sociality means that young birds can be guided through migration by older individuals, although there is evidence that inexperienced birds can get very lost if they become separated from their group (Lees and Gilroy 2022, 78). The adult birds in the numerous flocks I notice along the edge of the North Sea on this day presumably knew that the coast provided a clear and familiar line that could be tracked as they oriented southwards, as well as safe feeding and resting areas if the weather conditions got too bad or if some in their flock were tired after the overnight struggle through wind and rain as they crossed the North Sea. Their passage at this time would have emerged from both the weather and feeding conditions in the areas they had been staging in coastal Norway. The group would all need to be in good enough condition to have the energy to make a long flight over uninhabitable seas and they would look for helpful following winds and clear conditions to ease their flight. Like many birds, they might have some inkling when bad weather will hit, but they could still be surprised by unexpected changes in conditions that turn their passage into an ordeal amongst the weather-world (Ingold 2011, 126–135) and the limits of life.

Like many geese, they had 'traditional' wintering quarters on the Solway Firth in southwest Scotland, and I suppose many of those I saw passing would have made it there by the end of the day. But not all is stable in the world of Barnacle Geese because their migration is changing. For some time, research has demonstrated that migrating geese track a 'green wave' as they head towards their breeding grounds (van der Graaf et al. 2006). This wave is the emerging new growth of grass as the spring progresses. Geese feed on this nutritious growth as they migrate, storing up fat reserves in preparation for breeding.

To study these issues, scientists have tagged many Barnacle Geese with GPS transmitters to track their movements (Shariatinajafabadi et al. 2014). This has revealed that migrating geese gather the conditions of the places where they stop to feed into their bodies and lines of flight. Favourable or stressful conditions in one place have knock-on effects in other places where geese stop. The geese seem able to track the green wave responsively. Conservation clearly requires attendance to movements and responses of geese along their whole lines of flight, as they track the edges of inhabitability on their migrations.

The sorts of changes that are happening with Barnacle Geese have been replicated in many other wildfowl species in Europe. Russian breeders such as Bewick's Swans (Nuijten et al. 2020) and White-fronted Geese (Hearn 2004) that used to winter in large numbers in Britain now 'short-stop' further east, seemingly in response to milder winters. The birds only migrate as far as they need to escape the limits of uninhabitable ice and snow, although sometimes they still arrive in Britain in harder winters. The seasonal oscillation between green and white waves shapes the movements of wildfowl along clear and relatively simple limits. Destinations for winter might seem 'traditional', as they are used for many years by birds that are familiar with their ready inhabitability. But conditions can shift and the limits of life shift with them; destinations are shaped by these emerging limits rather than being pre-ordained final points towards which migration is automatically oriented.

Limit 2: Filling gaps

16 December2019. Hakodate, Hokkaido

I had arrived in Hakodate in southern Hokkaido the previous evening before meeting up with Japanese anthropologist Shiaki Kondo and expat British birder Stuart Price the next morning. Our first stop is near Hakodate docks, with Mount Hakodate bearing down upon us. There are no Brent Geese initially but then six arrive, settling close by. We walk closer and I am struck by how approachable they are. Soon, they come out of the water and start nibbling at seaweed.

These birds are from the small Pacific population of the Black Brant subspecies *nigricollis*. There is uncertainty as to where they breed, but around the Lena River delta in northern Siberia seems likely (Shimada et al. 2017). They pass through Hokkaido in autumn and smaller numbers winter. They are quite specialised, feeding mostly on coasts where eelgrass is abundant. Unlike Barnacles and some other Brents, they have not started feeding regularly on farmland, although the coastal destruction caused by the 2011 tsunami encouraged them inland for a time and away from the eelgrass (Shimada et al. 2018). We head east to the small town of Shinori and find a large group of Brents near the harbour. We count 47 and Stuart comments that they were absent when he had been there the previous weekend.

There was concern in Japan about the fortunes of Brent Geese, particularly as their favoured food of eelgrass was threatened and declining. The limits for such specialised birds can be more intensely felt, with fewer readily inhabitable places. The catastrophic conditions of the tsunami had provoked a creative response from the geese, as they moved inland but, thus far, there has been no longer-term shift in feeding and the birds have returned to their coastal eelgrass-rich lagoons. Should conditions change in the future and the limits of life challenge the geese again, experienced birds may remember the rice fields and the refuge they provided.

The wintering quarters of most Hokkaido Brents are also still uncertain. Scientists have recently started using satellite tags to follow them (Sawa et al. 2020). The most significant revelation came from the goose tagged Y18. After staging at Notsuke in eastern Hokkaido, Y18 flew across the Sea of Japan to arrive near Tanchon on the east coast of North Korea. It had long been suspected that this was where many Brents wintered, but it was hard to get data. Migratory birds like Brent Geese cross international boundaries, complicating both science and conservation. Their migrations are gatherings of different places into lines of flight and life. In doing so, they demonstrate that fixed political boundaries cannot contain the inherent in-betweenness of migration and that concerns for them need careful political work. The limits that stress our own relations need creatively shifting to shared responsibilities, while also acknowledging that political agreements and protections are contingent on the ever-shifting limits of movement in migration.

The commonest way in which conservation attempts to relate places of migration is through the concept of flyways. By understanding migrations as occurring along 'flyways', work can be done across international and geographical borders to manage habitats for the benefits of these species. Tracking birds as they track the lines of flight that emerge between the limits of life has enabled conservationists to know if, when, and how help might be needed and where. The risk for conservation, however, is that shifting political and ecological conditions can constrain intervention. Thinking of flyways as a set of points connected by migratory routes is always vulnerable to these changes, in which the points of protection are fixed and unable to move with the emergent in-betweenness of migration. Understanding migration as a wayfaring line of flight in which places are gathered in the body of the bird and its movement may not entirely resolve this problem, but it could help to shape different kinds of flyway and different forms of protection that are more responsive to shifting limits.

Limit 3: Changing direction

2 October 2016, Aberdeen, Scotland

It had been cold and clear overnight: perfect conditions for birds to migrate. I walked along the street where I live early in the morning, and it wasn't long before I realised what had happened. Some large sycamores loom over the street, and I hear a thin call coming from the canopy. It's a call I continue to hear throughout the morning. It comes from a bird I once thought of as a rarity in Britain. This was the day my view on this changed. The bird is a small, insectivorous species called a Yellow-browed Warbler. It spends the summer in the vast boreal forests of Siberia, breeding no closer than the Ural Mountains. I have seen them in the winter in Thailand, where that thin call seems to come from every bush. What, then, is a bird that breeds in Siberia and spends the winter in southeast Asia doing on a street in eastern Scotland in early October? This was not the first time I had seen them in Britain, but they were, supposedly, a vagrant species that was 'off-course'. By the end of that day, I had found at least ten in my area. That autumn, it was estimated that over a hundred were reported in the county of Aberdeenshire alone (Northeast Scotland Bird Club 2018). The national database logged over 5,000 records in Britain over the same period (BTO 2016). This didn't seem like a few random 'lost' birds but something more concerted. But why were they here? And where were they going?

The established vagrancy theory was that they arrived in northwest Europe in autumn because of a phenomenon called reverse migration (Lees and Gilroy 2022, 43–51). Migratory birds, it is argued, are born with a 'genetically-determined migratory direction' (Lees and Gilroy, 43), but sometimes, this is reversed, the bird mistakes north for south, and birds migrate 180 degrees the opposite way to normal. Yellow-browed Warblers normally migrate southeast from Siberia to the warm tropics of Asia. Some, however, migrate northwest. Through reverse migration theory, it is speculated that these birds, as well as other species with similar breeding and wintering ranges, continue into the Atlantic, hoping for welcoming tropical forests but instead succumbing to a watery grave. In this model, they are understood as faulty machines that are ultimately doomed.

But what if, rather than being an evolutionary dead end, these birds were exploring new possibilities for life? The orthodoxy of reverse migration was questioned by two researchers, James Gilroy and Alexander Lees (2003), Lees and Gilroy (2022), who argued that Yellow-browed Warblers and several other Siberian passerines migrated in various directions and the bias to northwest Europe was probably because of the large numbers of birdwatchers living there who would ensure that any birds that arrived were likely to be found. Instead, they argued that many Siberian birds were 'pseudo-vagrants' (Gilroy and Lees 2003, 433) that had, perhaps accidentally, discovered new wintering quarters in western Europe. Increasing numbers being recorded further south in Iberia were indicative of this shift. In the years since, Gilroy and Lees appear to have been vindicated, as warblers are being found more regularly in southwestern Europe and northwest Africa in winter (Lees and Gilroy 2022, 71).

The situation is still mysterious in part because Yellow-browed Warblers are resistant to scientists' methods of tracking migration. They are too small to have radio transmitters attached to them and can also be hard to find, particularly away from coastal areas where they tend to be most concentrated. But it seems likely that the warblers now regularly passing through western Europe in autumn are not simply the faulty machines they were once imagined to be.

Despite helping to develop new, less mechanistic, ways of understanding the trends in Yellow-browed Warbler migration, Gilroy and Lees still draw in their arguments about vagrancy on mainstream scientific theories of migration. A problem with these theories comes from the way they consider the movements of birds as a form navigation that is analogous to certain forms of human navigation. That comparisons are drawn with human navigation is not, itself, the problem. The problem is with the kinds of human movement that are used, as the following example illustrates:

> When humans think about navigation, we naturally conjure up maps of the world in our minds. This ability is not innate, however – we are fortunate that our predecessors have meticulously mapped the planet, providing us with visual resources that we can memorise and use to keep track of our location within the world around us. Our capacity for navigation thus depends almost entirely on learned information, much of which has been handed down from previous generations, as well as technological solutions like GPS – all of which are luxuries not afforded to most migratory birds.
>
> *(2022, 14)*

Although the authors contrast this kind of human navigation with how birds learn to migrate, the emphasis on human movement as technologically grounded in maps, compasses, clocks, and GPS is central to the analogies they and other scientists use to understand bird migration. For humans to 'know where they are', it is assumed they require a map (perhaps 'mental') and a compass. Although birds clearly lack the 'visual resources' of humans, it is assumed that they must have something similar internal to their bodies to 'know where they are'.

From the above, one might wonder how humans ever found their way around before the relatively recent development of maps and compasses, let alone GPS. But it is clear from the widespread literature on large-scale human movements that humans have been able to move around accurately and confidently without the use of any of these technologies at all, both on land and sea (Ingold 2000, 300). Although humans and birds clearly have different bodily capacities, both in terms of their ways of moving and of perception, how might explanations of bird migration look different if analogies are drawn with humans who are not modern, technologically equipped navigators?

The emphasis that scientific literature about bird migration places on 'navigation' is likely the root of the restrictive analogies drawn between bird and human movements. The anthropologist Tim Ingold (2000, 273–304) contrasts navigation with wayfinding (later wayfaring) as follows:

> To use a map is to navigate by means of it: that is, to plot a course from one *location* to another in *space*. Wayfinding, by contrast, is a matter of moving from one *place* to another in a *region*.
>
> *(2000, 274, emphasis in original)*

Ingold defines a 'region' as places that are 'nodes in a matrix of movement' (ibid.) and that these places have histories rather than locations. Ingold's argument is, in part, a critique of the widely held theory from cognitive psychology of 'mental maps', a theory that has also been used analogously to understand bird migration. According to Lees and Gilroy:

> Migratory birds are apparently able to develop 'mental maps' during their early years of life, using cues they experience during their first migratory journeys, to allow them to assess exactly where they are in relation to specific localities such as former breeding locations, stopover sites or winter territories.
>
> *(2022, 14)*

Following this way of thinking, what Ingold calls a 'complex structure metaphor' (2000, 274), migration is a relatively simple action of connecting movement to the complex structure of the mental map, rather like the moving position marker on the map of a satellite navigation device. Ingold argues instead for a 'complex process metaphor' (2000, 274–275). In this, there is much less requirement for pre-structured ideas, such as mental maps. The complexity lies in the skilled practice of wayfinding, in which the organism feels its way towards wherever it is heading, with ongoing perceptual monitoring and adjustments. As such, it is not about a correspondence between a complex, pre-existing mental map and worldly territory but the continual immersion of a creature in an unfolding world through complex, skilled perception and action. If we were to consider bird migration in these terms, as Ingold advocates for the wayfinding of many humans, could it help us understand the appearance of Yellow-browed Warblers in western Europe?

There are, however, some potential barriers for understanding Yellow-browed Warbler migration, and that of many other small passerines, in ways that emphasise 'skilled performance' and a fine tuning of perception. The first of these is that, like many passerines, Yellow-browed Warblers appear to migrate alone. Unlike the geese discussed above, they must find their own way, separated from experienced family members. Many of the Yellow-browed Warblers seen in Europe appear to be juvenile birds on their first migration. This was once considered a supporting factor for reverse migration theory, where juveniles could easily be explained as 'faulty machines' fresh off the production line that would become 'factory rejects' never returning to breed. But juvenile birds that migrate alone present various challenges for understanding bird migration. The most obvious is to explain how they know 'the right way' to go. This is where scientists have tended to assume innate inclinations that 'programme' birds to know which way north is and to fly in a particular orientation towards it. Lees and Gilroy describe this process thus:

It appears that to correctly develop the tools needed to robustly navigate on a first migration, a juvenile must go through two key steps:

- After hatching and prior to migration, a bird must develop a capacity to use at least one non-magnetic compass – either by memorising star maps… or developing a capacity to finely track the passage of the sun and/or pattern of polarisation through the day…
- Next, they must use these learned compass cues to calibrate their innate magnetic compass, ensuring they can use it to locate true north despite magnetic variability.

(2022, 28)

Here again, Ingold's complex structure metaphor is clear. The birds develop a detailed 'map' of both land and sky, and their movement is then a simple matter of flying in an innate direction in relation to this map. But is it possible for juvenile birds on their first migration to 'feel their way' around as part of a complex process that relies less on a mental map and a genetic pre-disposition that plots its movement against it? According to Ingold, at least for humans, 'we know *as* we go, not *before* we go' (2000, 288, emphasis in original). Can Yellow-browed Warblers, exploring new paths to western Europe, know as they go? Can their being and knowing be brought together in movement, rather than separated out into memorised maps and orientations, on the one hand, and the navigation of migration on the other?

While it is still impossible to determine exact answers to these questions, some progress can be made by considering certain assumptions about small passerines and their nocturnal, solitary migrations. The first is to consider whether these birds are 'alone' at all. Of course, they do not form tight, noisy flocks of familiar companions as geese do. But they move through a night sky when good conditions prevail and when, in many cases, they will share the air with numerous other birds moving in similar directions. As each bird moves, it no doubt has some awareness of others doing the same. Bird migration is now tracked by radar in some areas, with websites such as BirdCast showing live maps from North America indicating huge movements of millions of birds on some nights. The sky on a good night for migration can be a lively, inhabited place in which many birds are feeling their way in proximity to one another. The air itself is also not the uniform space that maps represent it to be. For birds that fly within it, the air is constantly differentiated, just as the land appears to be for humans as they walk (Ingold 2021, 70). The flows of air, with birds moving with it, often calling song lines through the sky as they go, creates the trace of a path, albeit one that must be remade every night. The throng of Yellow-browed Warblers that arrived that morning in Aberdeen must surely have felt their way along similar paths that were at least within earshot of others.

One might still ask whether the path in the sky they followed was in the 'right' direction. Here, an experiment by Paolo Maccagno is instructive. He decided to run north through the Shetland archipelago in northern Scotland. In this activity, he was inspired by Feldenkrais, a form of education in movement. Maccagno (2019, 110–111) describes the way this approach understands movement as both action and direction. Thus, 'running' and 'north' are not separate actions and orientations but are inseparable in movement. The 'forwardness' of the movement is specified by its 'north-ness'. Likewise, running is not simply 'movement in itself' but in relation to the direction in which it emerges, and 'north' is not simply a point read from a compass but a direction that develops through the movement of the runner. One might also speculate that the movement of birds in migration is not so much a response to a pre-existing, innate orientation but a specifying process that comes out of the forwardness of their movement, through the richly structured and inhabited region of the sky.

31 August 2020: North Ronaldsay, Orkney

The island of North Ronaldsay has become famous as a place to observe bird migration. I'd arrived the previous day and set out in easterly winds, ideal for bringing birds from northern Europe. On reaching a small patch of willows, I soon found a surprising but now familiar species in front of me. A Yellow-browed Warbler was quietly hopping about the bushes. There was something more significant happening here though. Although they are now regular in autumn, here was one in late August, weeks earlier than the normal first arrival. I later discovered this was the earliest ever autumn record in Britain. Was this a new limit of inhabitation being explored right in front of me? Could this have been a creative pioneer making its way across northern Europe much earlier, and perhaps from somewhere closer, than its congeners? I cannot know what challenges, or opportunities, had brought it here so early, but the lines of life it was generating seemed less pre-ordained than are often attributed to solo migratory birds; its migration could surely not be understood simply as an effect of a faulty orientation or timing. By considering this as emerging from experiences (both individual and collective) of the limits of life, I could instead see Yellow-browed Warblers as responsive *to* and responsible *in* the world.

Whatever the reasons, most agree that there is something unusual going on with Yellow-browed Warblers. There are rapid and surprising shifts and possible adaptations to changing conditions. But, although this is a species that appears to be 'doing well' in the sense of expanding where it can live and finding apparent solutions to the limits of a changing world, the volatility of these changes can trouble humans. In the autumn of 2021, after several years in which Yellow-browed Warblers were numerous along the east coast of Britain, hardly any appeared, with just one finding its way through the willow bushes in my local area. I and many others felt a twinge of what I term 'the anxious semiotics of the Anthropocene' (Whitehouse 2015) when confronted with this absence of a bird that had, quite rapidly, become a familiar and evocative feature of the autumn. It was a reminder that what we come to take for granted can quickly be undermined in an unpredictable world.

Limit 4: Adapt or die

29 December 2014. Pak Thale, Thailand

I've just arrived for a holiday. There are lots of birds I'm hoping to see, but one stands out in particular: the oddly named Spoon-billed Sandpiper. The bird is listed at number 37 in the ominously titled book '100 birds to see before you die' (Chandler and Couzens 2008), and I've wanted to see one as long as I could remember. I arrive mid-morning by some salt pans at Pak Thale, a couple of hours southwest of Bangkok. The pans are festooned with waders, most of them having made their way down the east Asian flyway from Siberia to winter here. Eventually, I find a Spoon-billed Sandpiper feeding distinctively, head down, filtering through water with its bill.

Why do Spoon-billed Sandpipers have such a hold over birders? They look like many other small shorebirds that scamper across the mudflats and beaches of the world. The strange spatulate bill is something unique though. It manages to be both familiar and utterly distinct. It's also now extremely rare. They breed in the Chukotka Peninsula in eastern Siberia and from there journey down the east Asian coast to spend the winter in southeast Asia. Never abundant, recent declines have left perhaps fewer than 300. I may have seen one before I die, but I wondered if the sandpiper might die out before I do.

12 January 2019. Minjiang Estuary, China

I headed out onto the mudflats near the city of Fuzhou, hoping to renew my acquaintance with Spoon-billed Sandpipers. As the tide rose, two soon appeared at close quarters, keeping their striking bills almost constantly underwater as they fed.

Scientists have recently started tracking Spoon-billed Sandpipers. At a recent webinar (Oriental Bird Club 2021), researcher Nigel Clark discussed the case of a sandpiper tagged HU. HU was migrating through eastern China when it took a striking diversion out into the South China Sea. He worried this was an effect of the tag on the bird's movements, but then checked the weather conditions. It seemed HU had encountered a cyclone and, rather than flying through it, had traced a curved path around its edge. The migrating bird was responsive to conditions, not tracking a pre-determined point-to-point route but, finding itself caught up in the fluctuations of the weather-world, tracking alongside the wall of the cyclone rather than being overwhelmed by passing through it.

FIGURE 3.2 Spoon-billed Sandpipers at Minjiang Estuary, Fuzhou, 12 January 2019.

Despite this illustration of the resilience of HU, Clark sounded a pessimistic tone. Spoon-billed Sandpipers are threatened by multiple factors: climate change, hunting, habitat loss, and pollution. He suggested that, even though the rate of decline is slowing, they could disappear by the 2030s. They have unique adaptations and creatively respond to the weather-world, but, like many other creatures, they are confronted by limits on their existence that humans are shaping. These new limits challenge them as to what path their lives can make. A storm can be responded to, in most cases, but other catastrophic problems are more challenging. Considering the world in terms of the possibilities of the limit and the limits of responsiveness might thus bring new understandings of how human action shapes what ordeals birds can and cannot respond to through migration.

The cases I discuss above merit comparison. In each, individual birds are able to migrate and to deal with the ordeals of the weather-world they encounter, at least so long as they have opportunities to rest and feed as they move. In the case of Barnacle Geese, the places where they do this are shifting, even if the journey unfolds along a similar path as it once did. Yellow-browed Warblers have developed new migrations in a different direction. It might be that this journey presents fewer challenges than it once did because winter conditions have become more benign so more birds migrating westwards are able to survive, perhaps by moving across the similar conditions of latitude rather than the diverse challenges of longitude

(Maccagno 2019, 101). These explorations become over-turnings in their move-ments rather than tragic dénouements. Spoon-billed Sandpipers can still cope responsively with the ordeals of migration, such as storms, but they find it harder to encounter safe places to rest and feed on migration and through the winter.

Creativity and catastrophe

In this chapter, I have attempted to reconsider ideas about bird migration. I do this not to refute scientific explanations but to reframe them in terms of wayfaring rather than navigation and through the idea of the limit as a creative ordeal that emerges from the movements of birds at the edges of inhabitability. I have argued that the concept of the limit helps to think through the ordeals of migration in three ways: by reframing migration as emerging from the organism-in-its-environment rather than as pre-existing organism and environment that are added together, by reconsidering the movements of migration as wayfaring rather than point-to-point navigation, and by thinking of migration not as a route between origin and destina-tion but as movement that is inherently 'in-between' at all times.

The case of Barnacle Geese helps to elucidate the indivisibility of organism and environment by demonstrating that their capacity to inhabit a place is influenced by conditions in other places they move through and is thus not simply a case of individual fitness being matched with external environmental conditions. Instead, 'inhabitability' is a quality that emerges from the interplay of geese and environment in motion. A place is 'inhabitable' not simply because of its intrinsic physical condi-tions but because of the lines of flight the geese have made, lines that draw different places and conditions together in their bodies and shape their interaction with the ever-shifting weather-world.

The shifts in migration in Yellow-browed Warblers show that orientation itself is not just a pre-existing inclination but can be considered as 'feeling the way' in movements that are simultaneously both action and direction. Their wayfaring through the night skies is not simply a solitary organism matching its position to a pre-existing mental map in order to reach a fixed point but is bound up in the mixtures of the weather-world and the movements of others making paths through the air in sound as well as movement.

The inherent 'in-betweenness' of migration is shown by Barnacle Geese tracking a 'green wave', an in-between seasonal flow that shifts the limits of inhabitability. Other birds that 'short-stop' in response to climate change suggest that migration is not point-to-point navigation but always 'in-between' in relation to the limits of inhabitability.

Migrating birds can deal with limits and can use them creatively and respon-sively. Some may prosper from the new possibilities that emerge, like the Barnacle Geese following shifting green waves or Yellow-browed Warblers exploring new possibilities in warming western Europe. Birds respond not only individually but also collectively, in flocks that share knowledge and experience or in populations that migrate in specific directions. But the limits of life that can be creative can also be catastrophic. For every storm that a Spoon-billed Sandpiper can move around, or every tsunami that presents Brent Geese with novel alternatives, there are conditions that are more challenging to survive where the capacity to respond to limits is impaired. In these cases, as Plumwood might put it, there is no creative revelation and no reprieve from the tensions of catastrophe.

Acknowledgements

I would like to thank Shiaki Kondo, Stuart Price, Yusuke Sawa, and David Anderson for assistance with the research on Brent Geese in Japan. That part of the research was funded through an ESRC and AHRC: UK-Japan SSH Connections grant.

References

British Trust for Ornithology. 2016. Record Autumn for Yellow-browed Warblers. https://www.bto.org/our-science/projects/birdtrack/2016-record-autumn-yellow-browed-warblers Last Accessed: 17/06/22

Chandler, David and Dominic Couzens. 2008. *100 Birds to See Before you Die*. London: Carlton Books.

Despret, Vinciane. 2016. *What Would Animals Say if we Asked the Right Questions?* Minneapolis: University of Minnesota Press.

Gilroy, James and Alexander Lees. 2003. Vagrancy Theories: Are Autumn Vagrants Really Reverse Migrants? *British Birds* 96: 427–438.

Hearn, Richard. 2004. *Greater White-fronted Goose* Anser albifrons albifrons *(Baltic-North Sea population) in Britain 1960/61–1999/2000*. Waterbird Review Series. Slimbridge: The Wildfowl and Wetlands Trust/Joint Nature Conservation Committee.

Ingold, Tim. 2000. *The Perception of the Environment: Essays in Livelihood, Dwelling and Skill*. London: Routledge.

Ingold, Tim. 2007. *Lines: A Brief History*. London: Routledge.

Ingold, Tim. 2011. *Being Alive: Essays on Movement, Knowledge and Description*. London: Routledge.

Ingold, Tim. 2021. *Correspondences*. Cambridge: Polity Press.

Lees, Alexander and James Gilroy. 2022. *Vagrancy in Birds*. London: Helm.

Maccagno, Paolo. 2019. Through these Walls: Steps to an Anthropology of the Limit. University of Aberdeen. Unpublished PhD thesis.

Newton, Ian. 2008. *The Migration Ecology of Birds*. London: Elsevier.

North-east Scotland Bird Club 2018. *North-east Scotland Bird Report, 2016*. North-east Scotland Bird Club.

Nuijten, Rascha, Kevin A. Wood, Trinus Haitjema, Eileen C. Rees, and Bart A. Nolet. 2020. Concurrent Shifts in Wintering Distribution and Phenology in Migratory Swans: Individual and Generational Effects. *Global Change Biology* 26: 4263–4275. https://doi.org/10.1111/gcb.15151

Oriental Bird Club. 2021. Spoon-billed Sandpiper: OBC Webinar. https://www.youtube.com/watch?v=293XGxD0J5A Last Accessed: 17/06/22

Plumwood, Val. 2012. *The Eye of the Crocodile*. Canberra: ANU E Press.

Sawa, Yusuke, Chieko Tamura, Toshio Ikeuchi, Kaoru Fujii, Aisa Ishioroshi, Tetsuo Shimada, Shirow Tatsuzawa, Xueqin Deng, Lei Cao, Hwajung Kim, and David Ward. 2020. Migration Routes and Population Status of the Brent Goose *Branta bernicla nigricans* Wintering in East Asia. *Wildfowl Special Issue 6*: 244–266.

Shariatinajafabadi, Mitra, Tiejun Wang, Andrew K. Skidmore, Albertus G. Toxopeus, Andrea Kölzsch, Bart A. Nolet, Klaus-Michael Exo, Larry Griffin, Julia Stahl, and David Cabot. 2014. Migratory Herbivorous Waterfowl Track Satellite-Derived Green Wave Index. *PLoS One* 9(9): e108331. https://doi.org/10.1371/journal.pone.0108331

Shimada, Tetsuo, Naoya Hijikata, Ken-ichi Tokita, Kiyoshi Uchida, Masayuki Kurechi, Hitoshi Suginome, and Hiroyoshi Higuchi. 2017. Spring Migration of Brent Geese Wintering in Japan Extends into Russian High Arctic. *Ornithological Science* 16: 159–162. https://doi.org/10.2326/osj.16.159

Shimada, Tetsuo, Yumi Yamada, Nayoa Hijikata, Ken-Ichi Tokita, Kiyoshi Uchida, Masayuki Kurechi, Hitoshi Suginome, Yasushi Suzuki, and Hiroyoshi Higuchi. 2018. Utilisation of Terrestrial Habitat by Black Brant *Branta bernicla nigricans* After the Great East Japan Earthquake of 2011. *Wildfowl* 68: 172–182.

van der Graaf, Sandra, Julia Stahl, Agata Klimkowska, Jan P. Bakker, and Rudolf H. Drent. 2006. Surfing on a Green Wave – How Plant Growth Drives Spring Migration in the Barnacle Goose *Branta leucopsis*. *Ardea* 94(3): 567–577.

Whitehouse, Andrew. 2015. Listening to Birds in the Anthropocene: The Anxious Semiotics of Sound in a Human-Dominated World. *Environmental Humanities* 6 (1): 53–71. https://doi.org/10.1215/22011919-3615898

4

HUMANS AND BIRDS ON BRITISH FARMS, 1950–2000

Paul Merchant

Introduction

Over the period 1950–2000, agricultural land continued to present the largest area for potential interaction between humans and wild birds in Britain. We know that certain species such as rooks and wood pigeons were regarded as threats to crops and deterred or controlled, and that other species were valued as game (Murton and Wright 1968, 117–209; Lovegrove 2007; Martin 2012). But we know surprisingly little about how farmers and farm workers viewed those other species of more or less common wild birds that make use of agricultural land, such as skylark, song thrush, buntings (corn, cirl and reed), curlew, peewit (lapwing) and yellowhammer (O'Connor and Shrubb 1986, 10–37). This is the ground this chapter covers, using oral testimony of some of the humans concerned.

The period 1950–2000 was one of significant change in British farming. Dramatic increases in production depended upon the amalgamation and speciali-sation of farms and widespread uptake of new practices and technologies, notably much greater use of artificial fertiliser and agrochemicals, increased autumn planting of cereals, more intensive management of grass, a switch from hay to silage and use of ever larger and faster field machines. From about 1960, ecologists began to study the effects on birds of these changes, producing new knowledge of the nesting and feeding behaviour of farmland birds and linking declines in prevalence to specific farming practices (Jenkins 1984; O'Connor and Shrubb 1986; Moore 1987). In the texts of these ecologists, farmers tend to be in the background, with just the occasional reference to their relative lack of interest in the birds in question: 'it is important to remember that farmers are interested in farming, i.e. the business of raising crops and stock to the best of their ability, and not in managing land for other purposes' (O'Connor and Shrubb 1986, 243); 'the vast majority of species ... are

DOI: 10.4324/9781003334767-6

largely ignored unless the farmer is a naturalist' (Moore 1987, 108). It was not until the 1990s that a significant number of these farmers were being invited to engage in agri-environment schemes that often featured certain birds as targets for conservation (Evans 1992; Winter 1996).

Inattention to birds as a focus of study

If farmers were inattentive to birds on their farms in the second half of the twentieth century, then – I suggest – so were many others. It may be true to say that birds are among plants and animals that there is a 'widely shared attraction to … in the United Kingdom' (Lorimer 2007, 917) or that 'by the early 1950s, birdwatching had achieved considerable cultural prominence' (Macdonald 2002, 56; Davis 2011, 226). The Royal Society for the Protection of Birds certainly had many more members than its equivalent organisation in Germany (Bargheer 2018, 4 and 11). However, the majority of people in Britain in the period were not members of ornithological organisations and, perhaps, not especially interested in birds (Evans 1992, 187; Bargheer 2018, 11). One reading of Guida's study of BBC broadcasts featuring birdsong after the First World War and during the Second World War is that these were engaging precisely because people did not usually listen closely to birds: the sound exceeded 'the everyday', moving people from the interwar 'humdrum' of, in the case of one correspondent, 'tragedy, politics, cricket and horse racing' or home front 'daily emotional tensions' (Guida 2018, 370–372 and 378). Moss is clear that though bird watching became 'more and more popular' after 1945 with 'new legions of birdwatchers', it was nevertheless the case that even by the 1970s only a 'minority … chose to spend their time watching birds' (Moss 2004, 191, 198 and 224). Van Dooren points similarly, but with a wider geographical scope, to a 'lack of popular interest' in the extinction of birds, prompting him to ask, 'Why do the last expressions of so many species leave this world unnoticed and un-mourned' except by a 'few conservationists'? (Van Dooren 2014, 140).

Though inattention to birds may well be widespread, it has not attracted scholarly interest. At least for Britain in the twentieth century, historical and sociological work has focused on the unusually close attention to birds of proponents of the 'new ornithology' of the interwar period, a wider 'new naturalism' post-war and of ornithologists of different kinds since (Matless 1998, 224–230). For example, Nixon reconstructs JA Baker's 'ten-year obsessive pursuit of wintering peregrine falcons in the farmland, valleys and coastal marshes of Essex' through his bird watching diaries; Toogood uses interviews with and published and unpublished writing of Max Nicholson to explore his exceptional attention to birds as a young man; and Lorimer follows surveyor 'Craig', someone who has gone to unusual lengths – driving down 'rutted tracks' in the dark for hours with the windows open – to find birds: 'Craig is passionate about counting corncrakes – in semi-retirement, it is the highlight of his summer' (Lorimer 2008, 390–392; Toogood 2011; Nixon 2017, 205). This chapter turns away from the unusually interested by considering life story interviews with farmers, farm workers and farm managers recorded for an oral history project that has no particular emphasis on wild birds.[1] In so doing, we find that inattention is itself an interesting focus for historical study of human–bird relations.

Oral histories of inattention to birds

Hamilton has argued that oral histories are a valuable but relatively untapped source for human–animal studies and shows that they can provide 'evidence' of 'day-to-day relationships' with birds that have left no other trace: 'Although the pigeons on Dolors's balcony and Carla's patio did not leave traditional historical records, the human testimony in these interviews provides evidence of their agency' (Hamilton 2018, 198–200). In this chapter, I too assume that my interview sources have what Thomson calls an 'empirical' content (Thomson 1999). Though the recordings are subject to all the pressures of narrative and personal 'composure' explored by oral historians over recent decades, they nevertheless refer to – and have value as evidence of – past presence, action, occurrence and experience (including inattention) (Abrams 2010; Merchant 2019).

The sources are oral history interviews of a particular kind: 'life story' interviews recorded over several sessions, between 7 and 12 hours long in total, including accounts of experience and action from early childhood. They allow us to consider what inattention to birds consisted of at particular times (stages of life and decades) and on particular farms. In attempting to draw out some of the richness of inattention, I have benefitted from readings of what Bonta and Protevi call 'Deleuze's ontology', in two main ways (Bonta and Protevi 2004, 14). One, DeLanda's development of assemblage theory emphasises the degree to which historical change does not require entities involved to be aware of their effects on each other, with all forms of organisation (from bodies to markets) being in some sense unintentional (DeLanda 1997, 2006). In the spaces and times considered in this chapter, agricultural change proceeded with little regard for birds, even though they were certainly involved. Two, Deleuze (and Deleuze and Guattari) recommend an apprehension of reality involving – at all scales and registers (ecological, social, geological, biological, cultural, etc.) – more or less durable stable states (hiding the activity which produced them), with opportunities for change and creativity in 'events' or 'lines of flight' (Massumi 1987, ix–xx; Bonta and Protevi 2004, 28, 106–107; Dodds 2011, 182–185). This ontology of capture and escape is echoed in interviewees' tendency to have – as I suggest in what follows – regarded birds as less tied down than themselves, relatively free to explore other 'solutions' to 'problems' of existence (Bonta and Protevi 2004, 23–26).

Farm birds in childhood

I repeat that the interviews were not recorded *in order* to study relations between humans and birds on farms. In many, no reference is made to wild birds, other than those that were targets for sport or control. Only a small number of interviewees say or suggest that they noticed or looked for other wild birds in childhood:

> When I first grew up, we always had hares, and we always had lapwing, and we always had curlew ... we used to have corncrake.[2]
>
> I got the *Observer's Book of British Birds* ... and so I got to know well, virtually every bird on the farm, I could name it.[3]
>
> Often we would go out ... looking for birds' nests ... there seemed to be more birds then than there are now. ... Lots of the songbirds, blackbirds, thrushes and so on ... It wouldn't be difficult for us to find a nest.[4]
>
> The first time that people remember that I was interested in birds is when I was told to go careful up the road, and I said, 'No, I'm, I'm not going along the road. I'm going along the hedge side to look for birds' nests.'[5]

These extracts suggest that birds were valued and appreciated but not in any way fostered or worried about. The third speaker, Steve Leniec, could readily 'stumble across' a nest as a child.[6] The impression is of a presence that was sufficiently abundant and reliable to be assumed. For two of the interviewees certain birds seem to have gained – by becoming less prevalent – a significance that they did not have at the time. They have become more notable through absence. Others have written along these lines through the idea of haunting (Parr 2010, 2; Tsing et al. 2017, 1–14).

Finally, the two interviewees who highlight the disappearance of certain birds since childhood are speaking of a local disappearance. Like most of the others we hear from in this chapter, they still live in the areas in which they grew up, sometimes on the same farm. The first speaker has been able to travel to hear corncrakes again:

> Oh the corncrakes ... I went specially to South Uist quite lately [recently] just to hear them. I had a week on South Uist. And luckily they are there in quantities.[7]

For all interviewees, there is left open the possibility that though certain birds are no longer *here*, they may be somewhere else. As we will see in the sections that follow – exploring interviewees at work on farms – an association of birds with mobility, relative detachment from the land, escape and opportunism can be bound up with inattention to them.

Prioritising work

For the majority of the interviewees, experience of farms as children was followed by experience of farms as farm workers, farmers or farm managers. Those who do not talk of interest in birds as children do not seem to have found themselves *more* affected by birds in the course of their working lives on farms of different kinds. Some give accounts of their careers that do not refer to wild birds at all. Only once, across all interviews, does a bird appear without the interviewer having asked about wildlife:

> The hill has a flat, completely flat top, before it drops down. Even to the point that top can be quite damp. I've seen snipe up there in wet weather. ... It was amusing to find the snipe up there.[8]

In those interviews where a question about engagement with wildlife is posed, these responses are typical:

INTERVIEWER: *What was your level of interest in wildlife as you were going about your work on the [farm]?*
Not a lot. [laughs] To be honest, no.[9]
INTERVIEWER: *Did you yourself notice any change in wildlife populations [on the farm]?*
I wasn't particularly interested, funnily enough ... just didn't feature at all. I regret that desperately now, because there must have been a heck of a lot of good ornithological interest ... but it passed me by.[10]

These responses, I suggest, point to both a tendency not to have been engaged by birds while at work on farms in the period and some pressure to express this now as a lack. For the first speaker, we might interpret the phrase 'to be honest' as indicating a general sense that inattention to wildlife may be viewed by the listener as negative. In the case of the second speaker – who developed an interest in birds through ornithological holidays in late career and retirement – lack of interest is also interpreted as a failing, though with the complication of personal reform and regret. In this case, he goes on to suggest that at the time, one kind of interest – farming – tended to leave no space for others; his son would report sightings of wildlife, including birds, 'but I, no, I wanted to know how much nitrogen to put on my winter wheat [laughs]'.[11] For these interviewees, farm work did not lead to attention to birds and focusing on agriculture could involve the overlooking or neglect of other kinds of experience.

Those interviewees who spoke of strong attention to birds in childhood suggest that they too tended to set this aside as they became farmers themselves. For example, the following account of the switch from hay to silage suggests a mode of decision-making that was strictly agricultural; the effects on birds were secondary to matters of productivity and viability:

> When I was a child we made hay, and you used to cut the hay at the beginning of June, and all the birds then had nested in the ... growing grass ... and got their babies off. ... But silage is always cut earlier, usually about the first week in May ... And because it came earlier in the year, sadly we now have no pewits, no curlew, and no corncrakes, entirely because of the change of farming practice. Which, I grieve about that really, but, that's what one had to do to make a living.[12]

Similarly, Poul Christensen – though 'fascinated' by farmland birds as a child – is clear that, when he started farming himself, work was prioritised. He took over the tenancy of Kingston Hill Farm in Oxfordshire in 1969:

INTERVIEWER: *I wondered whether, or to what extent you noticed, took interest in … the wildlife of the farm at the moment of taking it on?*
In truth, not a lot. All our energies were devoted to trying to turn this farm around from the condition that we found it in. … We were working flat out all the time just to do that, and so we didn't think about anything else really at the time.[13]

The farm had been used mainly as a shooting estate and was in 'appalling condition' with '350 acres of straw still lying in swathes on the fields and there was seventy acres of corn hadn't even been cut. …We ploughed day and night' (Figure 4.1).[14] Fences were put up, concrete roads and water piped laid, cow cubicles built and a milking parlour installed. Whatever birds might have been present tended to go unnoticed at this stage.

Birds reappear in his life story when he tells the story of, nearly ten years later, involving his now highly competitive dairy farm in the Countryside Commission's Demonstration Farms project. This was designed to show that nature conservation could be achieved relatively inexpensively, alongside conventional farming.

FIGURE 4.1 Farm worker chopping uncut corn at Kingston Hill Farm, November 1969. Credit: Poul Christensen. © British Library.

Representing the project in the farming and general press, Christensen stressed the seriousness and primacy of farm work over the accommodation of birds:

> Farming is my livelihood; wildlife, particularly birds, is my hobby. ... We are not farming conservationists. We are commercial farmers with an interest in conservation.[15]
>
> Our only course of action, if we really come under the screw, is to intensify still further ... We've got children; we're not philanthropists.[16]

We return to Poul Christensen and the Demonstration Farms project below.

Like Poul Christensen, Nicholas Watts was, he says, strongly interested in birds as a child: 'quite why I was so interested in it, I don't know'.[17] To the story of returning home by the hedge, reproduced above, he adds accounts of getting up early to record birds at a reservoir while the rest of his school slept and of disappointing conversations with his father who 'didn't know too much about birds'.[18] But when he started farming in the 1960s, he watched birds less, and when he did so, it was at sites away from the farm:

> I used to go, do a bit of birdwatching in the late Sixties and Seventies, but, I was pretty busy trying to build the farm up, you know. But, but interest was still there ... I was actually doing surveys in the Seventies for the Lincolnshire Bird Club and for the British Trust for Ornithology, but none of those, none of those surveys were on my farm.[19]

He did not, he says, concern himself with birds on and around the farm itself. Until the 1990s, he made decisions about cropping and the use of agrochemicals that were typical of commercial farmers of the period: 'In the 1960s our wheat crops were only sprayed with one herbicide, and sometimes an insecticide, and by the 1990s our wheat crops were being sprayed with three herbicides, two growth regulators, two insecticides, and three fungicides'.[20] A cost–benefit analysis was applied with no reference to possible effects on wild birds, even if these were later noted: 'When we had the winter barley, the corn buntings flocked round the winter barley, but as soon as we stopped growing winter barley ... the corn buntings were gone'.[21]

It was not until the early 1980s that Watts started paying more attention to birds on his farm. He began to take photographs of them and embarked on a survey during the nesting season. It is not clear what caused this change, but he says it was not a concern about prevalence:

> I was doing bird surveys ... but none of those surveys were on my farm. And I thought I would like to see what was on my farm.
>
> INTERVIEWER: *were you recording in order to monitor a decline that you thought was there, or were you recording just to record at that point?*
>
> Oh, just recording to record, yes.[22]

We return to Watts' surveys on his farm in the next section. But first I make three comments on the lack of attention to wild birds that has been explored in this section. One, farms were not understood as the right places for observing birds. We might like to draw in Raymond Williams here and suggest that the landscape of farms continued to be one in which only certain kinds of attention were prioritised by those working in it (Williams 1985, 120–126; Giblett 2012). Even if individuals – as children or adults – particularly enjoyed watching birds, they tended not to regard time spent at work on farms as opportunities to pursue this. Secondary sources suggest that this view of farms as inappropriate places for the observation of birds was shared by ecologists and ornithologists in the decades following the Second World War. Matless, for example, has shown that post-war 'new naturalism' depended, in part, on a separation of industrial agriculture on the one hand and reserved nature on the other (Matless 1998, 226–227). Which takes us to point two – also evidenced in this section – that farms tended not to be regarded as the right places for birds themselves. I suggest that it was amusing for Pippa Woods to find a snipe on a part of a field that didn't drain well because the bird was felt to be there by accident; it was a curious, whimsical presence.

I argue further that points one and two are linked by the assumption that birds can always be somewhere else. They are mobile and there are other places for them: other parts of the country with less intensive agriculture, nature reserves, national parks. Their opportunity for escape might have seemed, to farmers themselves, to exceed their own ability to escape commercial pressures to modernise or improve. As DeLanda shows particularly clearly, it continued to be the 'anti-market' processors and investors in agriculture, rather than the farmers, who could break free of the grounded limits of growing commodities (DeLanda 1997, 49, 129, 166 and 265–266). We might forgive interviewees for feeling that birds' opportunities for exploring 'lines of flight' were surely more promising than the average British farmer of the period, relatively fixed into conventional ways of improving their practice to stay profitable.

Engaging in schemes and surveys

From the 1960s, Britain's state-funded Nature Conservancy devoted more of its resources to research and lobbying with respect of the wider farmed countryside beyond nature reserves, with the establishment of its Monks Wood Experimental Station and Toxic Chemicals and Wildlife Section key developments (Moore 1987, 150–152; Toogood 2008). A Countryside Commission was formed in 1969, taking over the National Parks Authority, with authority over the 'wider countryside' (Evans 1992, 110). In 1970, a small organisation called the Farming and Wildlife Advisory Group (FWAG) was established to advise farmers on nature conservation measures; by 1987, it had 30 advisers (Moore 1987, 110; Winter 1996, 240–246). The Wildlife and Countryside Act of 1981 aimed to strengthen protection for Sites of Special Scientific Interest (SSSIs), and there was a gradual 'greening' of national and European agricultural policy in the 1980s and 1990s, reflected in agri-environment schemes, notably those associated with the designation of Environmentally Sensitive Areas (from 1987, covering 15 percent of farmland by 1994) and the Countryside Stewardship Scheme (from 1991, covering 91,937 ha by 1995) (Winter 1996, 208–234). Given these changes, by the 1980s, most farmers in Britain were likely to have been aware of growing public and political concern over the effects of farming practices on wildlife, including birds, and – especially from the 1990s – they have been able to engage in agri-environment schemes. Some of these schemes included, or even focused on, those incidental farmland birds we are concerned with here. In this section, I want to argue that involvement in such schemes was possible with no particular, or new, or renewed, attention to birds.

As we saw in the previous section, Poul Christensen – with his farm now 'a dairy farm, I think to be proud of' – decided in 1979 to allow the farm to be the first in the Countryside Commission's Demonstration Farms project, concerned with demonstrating relatively simple farm management practices for nature conservation.[23] The first step was a survey of the farm by external experts, including ornithologists:

> Anybody who had a land use interest came and looked at the farm. … There was the footpath people, there was the slugs and snails people, the bird people, the landscape people, the reptile [people].[24]

The 'bird people' – British Trust for Ornithology (BTO) surveyors – returned to the farm each breeding season to complete a two-sided sheet entitled 'BTO Farmland Common Bird Census' noting numbers of a list of species present.[25] Newspaper and magazine articles of the period featuring Kingston Hill Farm Demonstration Farm report the presence of '65 species of birds' as shorthand for success in nature conservation in general.[26] During visits from the BTO surveyors each year, Christensen had the opportunity to compare his own attention to birds with that of the surveyors:

> I'd occasionally walk round with some of these people and they'd hear just a snippet of birdsong and they'd write something down. And I said, "what have you written down?" And they'd say, "oh that's a … whatever it was". And I hadn't even heard it, or hardly heard it. And yet they knew precisely what it was.[27]

What I want to highlight here is that by engaging in the scheme, it was not necessary for Christensen himself to attend anew to birds directly. Instead, he had to arrange for the planting of trees, the cutting of hedges in certain ways at certain times, leaving of margins round fields, avoidance of stray fertiliser. Though involving a significant, voluntary and – at the time – unusual extra commitment of time and thought, this could all be done without having any more to do with birds than he did anyway. The hedges, for example, were cut by a contractor who supplied his own sight records:

> We don't even know when he's going to come; he just turns up and cuts them to the right shape and so on. ... On the back of the bill he sends us will be a list of the birds he saw and where he saw them. ... Real interest in it.[28]

As the availability of agri-environment schemes increased, other farmers – less pioneering than Christensen – became involved. They way in which they speak about these schemes does not suggest very much in the way of direct entanglement with birds:

INTERVIEWER: *To what extent was your own farm surveyed or advised by FWAG?*
It was surveyed and we did two or three schemes with them.[29]

> We have embarked on a Countryside Stewardship Scheme ... but as for sort of songbirds and things like that, we never had a survey done of the birds, farmland birds as such. [pause] No, I'm wrong. I think we did have a survey done ... now I think about that. When it was done, and by whom, and what the ... outcome was, I really can't remember.[30]

Pippa Woods speaks of the involvement of her farm in a cirl bunting scheme similarly. It was possible to be in the scheme by agreeing not to cultivate certain fields and to instead mow them to prevent succession of gorse in return for a subsidy. No new direct attention to birds was necessary:

> Great excitement about the cirl bunting suddenly cropped up and ... people who were interested in the cirl bunting sort of suggested that they must be here and that we should be in the cirl bunting scheme, so we said, fair enough. So we're in the cirl bunting scheme. David [interviewee's son] claims to have seen one ... I've never seen a cirl bunting here.[31]

I do not present these extracts as evidence of carelessness. These farmers entered into schemes in partnership with others who were tasked with attending to birds on their behalf. It was not necessary for interviewees to see or hear birds themselves; they just needed to follow certain rules on farming practices and timings. In this way, the schemes, however positive, may have had the unintended consequence of replacing whatever attention to birds farmers did exercise. In Christensen's interview, in particular, there is some evidence of a wish to defend the value of his own general 'fascination' with birds alongside impressive feats performed by trained surveyors:

> I pale into insignificance in terms of these people who come along and do these surveys ... I was in awe, I am in awe of their ability to identify birds and so on. But my interest in birds goes right back as a kid, because I was always fascinated by birds on the farm where I was brought up in Sussex. [32]

We return to the value of different kinds of engagement with birds in the conclusion. But first, I want to consider a different case: Nicholas Watts' own surveys of birds on his farm. Like others, he entered into agri-environment schemes from the 1990s (he mentions Countryside Stewardship) and had visits from BTO surveyors. However, he stands out in the collection of interviews as a farmer who *himself* surveyed birds on his own and neighbouring farms. Starting in 1981, twice in May and once in June before the farming day, he walked up one side of the farm and down the other side and plotted the locations of nesting birds on a map by listening and looking: 'most of the birds I would hear rather than see, or hear them first and then see them'.[33] Because he was walking the boundary of his farm, he could include birds on the farms either side. After the first ten years of doing this, he started to plot the results in a comparative way, and this process showed up the extent of declines in presence: 'It was one day in 1992 when I worked things out, and, found there was such a big difference'.[34] He also plotted the results for each year on cropping maps (Figure 4.2). Crucially, this highlighted the relations between the location of birds and field-scale land use. Of such a map featuring corn buntings and skylarks in 1982 and 1992, he explains:

> And, this is, this orange, this mauve-y colour is winter barley. The dots are singing corn buntings. And all the singing corn buntings are within reach of the winter barley crops, except this, this one here.

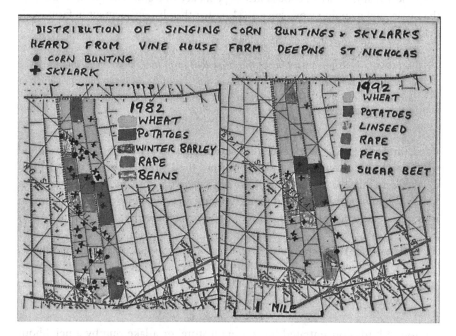

FIGURE 4.2 Nicholas Watts' map of corn buntings and skylarks on Vine House Farm and neighbouring farms in 1982 and 1992. Credit: Nicholas Watts. © British Library.

INTERVIEWER: Yes, sort of, middle on the right…. About halfway up on your neighbour's field.
Yes. Well, I don't know why he chose that, but anyway, he perhaps thought there wasn't room for him just up here, you know; there's four of them there isn't there. And, that's how it was in those days, in 1982. But by 1992, there's only one corn bunting left, we've stopped growing winter barley, and, there's a sixty per cent decline in skylarks as well.[35]

I suggest that through his survey and mapping work, Watts discovered the limits of birds' mobility. The corn buntings appear in the extract above as strongly constrained. Speaking of other plots too, what he highlights is the discovery of birds as grounded in two ways. One, they can't just be anywhere; options for landing and nesting are limited. Two, once nested, the birds Watts is concerned with 'really don't go more than 300 yards away from where they nest'.[36] They rely on short-range flight to a patchwork of horizontally arrayed food available at different times:

> You know, they really want to be, just here, where there's potatoes, there's, that one is, peas, and there's wheat, you know, they want to be, they want to have three or four. … When, when there's food in one crop, another one's run out of food, and so on you see. So, yeah, they've got the choice where to go.[37]

He is clear that he learned about birds' dependence of features of the farm through the survey and mapping work since 1981, rather than by engagement in farm work beforehand.

INTERVIEWER: Your understanding of this behaviour in relation to the fields and the space of the farm, did that come from plotting, or did you, did you know this already? Was this revealed to you by this sort of effort?
It was revealed by this, yes.[38]

From the early 1990s, he extended his surveying to cover a 400-acre area around the farm: 'I walk about ninety miles from April the 1st to June the 1st, up and down the dykes, in this area. … And then I can see, where the birds are, and where they're not'.[39] When he speaks with and of the resulting maps, what he draws our attention to is further discoveries of birds' reliance on particular ground conditions:

> So this area here stands out, here, because these are small farmers, and there's a line, electric line goes down here, and they like to sit up well and sing from that electric line. But those small farmers are not, haven't embraced all the new ideas that the bigger farmers have.[40]

He found birds concentrated 'near a grain store' or a lake dug by a neighbour for fishing.[41] These discoveries, he stresses, were new to him. They were a surprise, even though he had lived and worked in the area for 50 years: 'Through my survey … I realised what was missing, and I could see where the best, the best wildlife habitat was'.[42] He set out making changes on his farm in light of his new knowledge and continued to be *surprised* by birds' dependence on the ground. For example, he persuaded the local drainage board to lower the water level in a main drain only when necessary, rather than routinely, and he 'just couldn't believe that it would make that much difference'.[43] By telling us about his surprise at the discovery of birds' dependence on ground features, I suggest that Watts is also telling us something about the nature of his relative inattention to birds on the farm earlier in his career. He didn't just discover what birds liked; he was surprised by their lack of mobility. Like others discussed above, he had allowed himself to believe that birds were relatively independent of whatever was on the ground or close at hand. They were less free and vertical than perhaps he had imagined. Following DeLanda's reading of Deleuze, we could say that Watts discovered and was surprised by the 'contingently obligatory' relations between wild birds and farmland in the farm-bird 'assemblage' (DeLanda 2006, 12).

Conclusion

With the exception of land under agri-environment schemes in more recent decades, for much of the period covered by this chapter, the farmland available to birds was produced with little regard for them. Staying with DeLanda and assemblages for a moment, we could say that the farm-bird assemblage before (recalled in childhood memories) and after intensifications of the 1950s onwards were both the 'unintended consequence of intentional action … a kind of statistical result' (DeLanda 2006, 25 and 41). At the beginning of the period and before, farms were farmed in particular ways and certain birds were unintentionally supported. When cropping patterns changed and more and more effective herbicides were used to exclude weeds, some of the same birds were – just as unintentionally – not supported. Hay fields were made with as little regard for wild birds as silage fields.

From the 1980s, but not on a very large scale until the 1990s and beyond, agri-environment schemes have been successful in maintaining or producing surfaces (hedges, unsprayed areas, beetle banks, field headlands and margins) on which certain birds can land, nest and feed. The interviews suggest that these could be entered into by farmers without entering into new kinds of direct attention to birds. A key criticism of agri-environment schemes is that they are designed in accordance with a dominant, authoritative scientific way of knowing and may replace more spontaneous, individual kinds of environmental knowledge (Maskit 2008, 461–484; Lorimer 2015, 94–96). There is some evidence in the recordings that farmers could doubt their own noting of birds in the light of more expert attention. This is one way of reading parts of Poul Christensen's interview discussed above, and this hesitant response in particular:

INTERVIEWER: *And is birdsong on the farm something … that you've caught yourself noticing as you've been farming?*

Yeah. Yes, yes, you do, erm [pause] Oh yes, absolutely you do. I mean, why am I hesitating on that? When you get up to milk at 5 o'clock in the morning in May, it's just getting light and they're going at it hammer and tongs singing, the birdsong, really loud, a whole range of birds, and that's when I couldn't tell you [which birds]. I know some of the blackbirds and the robins and so on.[44]

However, far from suggesting that agri-environment schemes displaced a valuable, indigenous human–bird entanglement, the interviews considered in this chapter tend to suggest that such schemes were introduced to a farming community that was not especially attentive to or engaged by wild birds most of the time.

The chapter has shown further that inattention to birds on farms in the period has a content that can be explored in personal testimony. By listening carefully to oral history interviews, it has been possible to suggest that inattention was bound up with assumptions about the freedom and mobility of birds afforded by flight. In particular, farmers and other farm workers seem to have imagined birds to be – relative to themselves – free from the pressures of making a living from any particular bit of ground. We might speculate that oral history interviews could be used to explore the content of other imaginings or habits of thought that have played into our planetary plight (Bateson 1972; Guattari 2008; Dodds 2011; Latour 2018).

Notes

1 An Oral History of Farming, Land Management and Conservation in Postwar Britain, led by National Life Stories at the British Library, supported by the Arcadia Fund. In this respect, the interviews differ from the 'stories' collected by the Listening to Birds project, discussed in Whitehouse 2015, 63–66.
2 Interview with Jill Hutchinson-Smith, 2019, C1828/19 Track 4, British Library Sound Archive.
3 Interview with Poul Christensen 2019, C1828/08 Track 1, British Library Sound Archive.
4 Interview with Steve Leniec 2020, C1828/27 Track 1, British Library Sound Archive.
5 Interview with Nicholas Watts 2019, C1828/14 Track 2, British Library Sound Archive.
6 C1828/27 Track 1.
7 C1828/19 Track 4
8 Interview with Pippa Woods 2019, C1828/17 Track 4, British Library Sound Archive.
9 Interview with Nigel Young 2019, C1828/12 Track 8, British Library Sound Archive.
10 Interview with Robert Hart 2019-2021, C1828/13 Track 10, British Library Sound Archive.
11 C1828/13 Track 10.
12 C1828/19 Track 4.
13 C1828/08 Track 4.
14 C1828/08 Track 3.
15 Quote from Harvey, Graham. 1976. "Neighbours with Nature." *Farmers Weekly*, December 31, 1976, p. 57.
16 Quote from Winsor, Diana. 1982. "Enjoy or Destroy: A Rural Conflict." *Telegraph Sunday Magazine* 292, May 1982, p. 25.
17 C1828/14 Track 2.
18 C1828/14 Track 2.
19 C1828/14 Track 3.

20 C1828/14 Track 13.
21 C1828/14 Track 6.
22 C1828/14 Track 6.
23 C1828/08 Track 5.
24 C1828/08 Track 5.
25 Between 50 and 60 species are identified each year to 1992. Records courtesy of Poul Christensen.
26 For example, Davies, Gareth Huw. 1982. "And on that Farm…He has Badgers and Weasels – and 65 Species of Bird." *The Sunday Times*, August 15, 1982, p. 14.
27 C1828/08 Track 5.
28 C1828/08 Track 18.
29 Interview with Nigel Miller 2021, C1828/29 Track 3, British Library Sound Archive.
30 Interview with Tim Ruggles-Brise 2021, C1828/36 Track 6, British Library Sound Archive.
31 C1828/17 Track 11.
32 C1828/08 Track 17.
33 C1828/14 Track 6 and 7.
34 C1828/14 Track 6.
35 C1828/14 Track 9.
36 C1828/14 Track 9.
37 C1828/14 Track 9.
38 C1828/14 Track 9.
39 C1828/14 Track 12.
40 C1828/14 Track 9.
41 C1828/14 Track 9.
42 C1828/14 Track 11.
43 C1828/14 Track 12.
44 C1828/08 Track 17.

Bibliography

Abrams, Lynn. 2010. *Oral History Theory*. London: Routledge.

Bargheer, Stefan. 2018. *Moral Entanglements: Conserving Birds in Britain and Germany*. London: The University of Chicago Press.

Bateson, Gregory. 1972. *Steps to an Ecology of Mind*. London: The University of Chicago Press.

Bonta, Mark and John Protevi. 2004. *Deleuze and Geophilosophy: A Guide and Glossary*. Edinburgh: Edinburgh University Press.

Davis, Sophia. 2011. "Militarised Natural History: Tales of the Avocet's Return to Postwar Britain." *Studies in History and Philosophy of Biological and Biomedical Sciences* 42 (2): 226–232. https://doi.org/10.1016/j.shpsc.2010.11.027

DeLanda, Manuel. 1997. *A Thousand Years of Nonlinear History*. New York: Swerve.

DeLanda, Manuel. 2006. *A New Philosophy of Society: Assemblage Theory and Social Complexity*. London: Bloomsbury.

Dodds, Joseph. 2011. *Psychoanalysis and Ecology at the Edge of Chaos: Complexity Theory, Deleuze/Guattari and Psychoanalysis for a Climate in Crisis*. London: Routledge.

Evans, David. 1992. *A History of Nature Conservation in Britain*. London: Routledge.

Giblett, Rob. 2012. "Nature is Ordinary Too." *Cultural Studies* 26 (6): 922–933.

Guattari, Felix. 2008. *The Three Ecologies*. London: Continuum.

Guida, Michael. 2018. "Surviving Twentieth-Century Modernity: Birdsong and emotions in Britain." In *The Routledge Companion to Animal-Human History*, edited by Hilda Kean and Philip Howell, 367–389. London: Routledge.

Hamilton, Carrie. 2018. "Animal Stories and Oral history: Witnessing and Mourning Across the Species Divide." *Oral History Review* 45 (2): 193–210. https://doi.org/10.1093/ohr/ohy053

Jenkins, David, ed. 1984. *Agriculture and the Environment*. Cambridge: Institute of Terrestrial Ecology.

Latour, Bruno. 2018. *Down to Earth: Politics in the New Climatic Regime*. Cambridge: Polity.

Lorimer, Jamie. 2007. "Nonhuman Charisma." *Environment and Planning D: Society and Space* 25 (5): 911–932. https://doi.org/10.1068/d71j

Lorimer, Jamie. 2008. "Counting Corncrakes: The Affective Science of the UK Corncrake Census." *Social Studies of Science* 38 (3): 377–405. https://doi.org/10.1177/0306312707084396

Lorimer, Jamie. 2015. *Wildlife in the Anthropocene: Conservation after Nature*. London: University of Minnesota Press.

Lovegrove, David. 2007. *Silent Fields: The Long Decline of a Nation's Wildlife*. Oxford: Oxford University Press.

Macdonald, Helen. 2002. "'What Makes You a Scientist is the Way You Look at Things': Ornithology and the Observer 1930–1955." *Studies in the History and Philosophy of Biological and Biomedical Sciences* 33 (1): 53–77. https://doi.org/10.1016/S1369-8486(01)00034-6

Martin, John. 2012. "The Transformation of Lowland Game Shooting in England and Wales in the Twentieth Century: The Neglected Metamorphosis." *The International Journal of the History of Sport* 29 (8): 1141–1158. https://doi.org/10.1080/09523367.2012.690226

Maskit, Jonathan. 2008. "Something Wild? Deleuze and Guattari, Wilderness, and Purity." In *The Wilderness Debate Rages On: Continuing the Great New Wilderness Debate*, edited by Michael P. Nelson and J. Baird Callicott, 461–484. London: The University of Georgia Press.

Massumi, Brian. 1987. "Translator's Foreword: Pleasures of Philosophy." In *A Thousand Plateaus: Capitalism and Schizophrenia*, edited by Gilles Deleuze and Felix Guattari, ix–xx. London: Continuum.

Matless, David. 1998. *Landscape and Englishness*. London: Reaktion Books.

Merchant, Paul. 2019. "What Oral Historians and Historians of Science Can Learn from Each Other." *The British Journal for the History of Science* 52 (4): 673–688. https://doi.org/10.1017/S0007087419000517

Moore, Norman. 1987. *The Bird of Time: The Science and Politics of Nature Conservation*. Cambridge: Cambridge University Press.

Moss, Stephen. 2004. *A Bird in the Bush: A Social History of Birdwatching*. London: Autumn Press.

Murton, R.K. and E.N. Wright. 1968. *The Problems of Birds as Pests*. London: Academic Press.

Nixon, Sean. 2017. "Vanishing Peregrines: J.A. Baker, Environmental Crisis and Bird-Centred Cultures of Nature, 1954–73." *Rural History* 28 (2): 205–226. https://doi.org/10.1017/S0956793317000115

O'Connor, Raymond J. and Michael Shrubb. 1986. *Farming and Birds*. Cambridge: Cambridge University Press.

Parr, Joy. 2010. *Sensing Changes: Technologies, Environment, and the Everyday, 1953–2003*. Vancouver: UBC Press.

Thomson, Alistair. 1999. "Making the Most of Memories: the Empirical and Subjective Value of Oral History." *Transactions of the Royal Historical Society* 9: 291–301. https://doi.org/10.2307/3679406

Toogood, Mark. 2008. "Beyond 'the Toad Beneath the Harrow': Geographies of Ecological Science, 1959–1965." *Journal of Historical Geography* 34 (1): 118–137. https://doi.org/10.1016/j.jhg.2006.12.002

Toogood, Mark. 2011. "Modern Observations: New ornithology and the Science of Ourselves, 1920–1940." *Journal of Historical Geography* 37 (3): 348–357. https://doi.org/10.1016/j.jhg.2010.11.002

Tsing, Anna Lowenhaupt, Nils Bubandt, Elaine Gan, and Heather Anne Swanson. 2017. *Arts of Living on a Damaged Planet: Ghosts and Monsters of the Anthropocene.* Minneapolis: University of Minnesota Press.

Van Dooren, Thom. 2014. *Flight Ways: Life and Loss at the Edge of Extinction.* New York: Columbia University Press.

Whitehouse, Andrew. 2015. "Listening to Birds in the Anthropocene: The Anxious Semiotics of Sound in a Human-Dominated Worlds." *Environmental Humanities* 6 (1): 53–71. https://doi.org/10.1215/22011919-3615898

Williams, Raymond. 1985. *The Country and the City.* London: The Hogarth Press.

Winter, Michael. 1996. *Rural Politics: Policies for Agriculture, Forestry and the Environment.* London: Routledge.

PART II

Making sense of shared space

5

AIRBORNE

Experience and atmospheric movements in falconry practice

Sara Asu Schroer

In the essay "Hawks in Air", nature writer Richard Jefferies (1843) describes his observation of two birds of prey flying on motionless outstretched wings, rising higher and higher, seemingly defying the gravitational pull of the earth. It is easy to visualise him standing firmly on the ground, his head tipped back to better watch the two birds, observing them intently. He is fascinated by the apparent effortlessness of the flying birds and wonders how it might be possible, arguing about this problem at length without finding a satisfactory solution. Jefferies describes how the two birds started their flight above the treetops and while circling in broad spirals, without any visible exertions, rose higher and higher to disappear out of sight in the blue sky above. Without an understanding of what was going on in the airy spaces above, he was unable to interpret what was happening before his eyes – the rising air currents remained imperceptible to him.

Jefferies' observation and marvel at the birds' flight reminded me of my own, admittedly rather ignorant position, at the beginning of my fieldwork into the practice of falconry. Like Jefferies, I was intrigued by the beauty and elegance of soaring birds but was not yet familiar with the complex movements of air currents and weather that constitute the environments of airborne creatures such as birds of prey. In this chapter, I aim to show that working with falconers and their falcons, hawks, and eagles required an opening up of the earthbound perspective of conventionally "grounded" fieldwork, and I will consider the aerial world of birds of prey in which weather and air currents play a crucial role. I argue that through their daily intimate engagement with birds of prey in shared tasks, falconers get bodily and sensually attuned not only to their birds but also to the environments in which they are immersed. This draws their attention to the atmospheric constitution of the world, or the "weather-world" in Tim Ingold's terms, revealing a world in constant formation in which the weather has a constitutive influence on human and nonhuman movements and lifeworlds. This will be done through explorative

DOI: 10.4324/9781003334767-8

ethnographic stories aiming at drawing attention to the texturality of air, revealed in the engagement between falconer, bird, and the weathering environment (see also Azevedo and Schroer 2016). Considering material from my fieldwork, this chapter gives an insight into this aerial perspective of the world.

Hunting with birds of prey

Through a cooperative hunting relationship, falconers gain an intricate knowledge of how birds of prey use the air and develop an intimate understanding of how air currents direct the flights of birds, prey species, and consequently their own movements on the ground. Falconry is a hunting practice in which humans and birds of prey learn to hunt in cooperation with each other. This hunting cooperation is formed through a variety of taming and training methods and techniques particular to falconry and depends on a fine balance of wildness and tameness, dependence and independence of the falconry bird (Schroer 2015). Falconers understand their relationship with birds of prey as being based upon trust and a sense of companionship rather than as coercive and utilitarian. In order to train a falcon for cooperation in hunting, the bird needs to be "lured" and "politely charmed" into a relationship with the falconer and cannot be forced or bullied into submission. In order to do so, it becomes central for falconers to develop "a feeling for birds of prey" and to learn to understand the world in relation to these airborne creatures, who in so many ways perceive and act upon the world differently from the earthbound falconers (Schroer 2018).

In my ethnographic research, predominantly conducted not only in the various places in the British countryside but also including shorter fieldtrips to Italy and Germany, I became particularly interested in how falconers come to know and perceive their environments through the intimate cooperation with an airborne creature, who in so many ways perceives and acts upon the world differently to the earthbound falconer. More specifically, what kind of sense of communication had to be established in order for this particular kind of cooperation to work? I therefore paid particular attention to the learning practices in which both humans and birds are involved. These practices, which include different stages of taming, training, and hunting with birds of prey, are particular to falconry and involve a gradual process of familiarisation in which the falcon, hawk, or eagle is gradually accustomed to humans and their associated world (comprising sounds, movements, tools, infrastructure, other domestic and nondomestic animals, etc.). Falconer John described the beginning of this familiarisation process as such:

> When you have a new bird you need to understand that this bird has absolutely no idea what is going on, up until now she was flying about with other birds in an aviary only seeing humans from time to time without actually being handled much, and now she sits there on your fist, threatened, and without any idea of what to expect.

The aim of the first weeks in the falconer's care is directed towards overcoming this initial fear of the bird and crucially to establish a "common ground" of communication through which bird and human gradually learn to share a repertoire of meaningful relationships. Falconers described this process of familiarisation as a gradual attunement of the birds' senses to the unfamiliar surroundings, having to take into account their particular temporal and spatial experience in the way they move and behave towards them. Their visual experience of the world, for instance, allows falcons, as compared to humans, much more acute and fast responsiveness to what is happening around them. Allowing them to perceive movements that remain unrecognized by their human caretakers can help falcons adapt to their new environment and establish a stronger bond with their falconer.

The techniques adopted in the familiarisation process structure the bird's perceptual engagement with the world and influence it in ways that allow certain aspects to be revealed to the novice bird while others remain hidden. These techniques work through gradually orienting the bird's attention away from sights that might threaten her to more pleasant activities. Feeding especially plays an important role in the creation of positive experiences and expectations associated with the presence of the falconer. This active and structured influencing of the birds' perception is aimed at making the bird comfortable in this new environment and hence is a precondition for the learning processes that are involved in the subsequent training and hunting practices.

This process of familiarisation, however, is not one-sided, as falconers routinely underlined the importance of being able to attune to and open up towards the particular sensory and perceptual abilities of the bird they are working with. Working with birds of prey in shared tasks, both in regard to concrete learning events as well as lifelong learning experiences, is often described as a transformative process. With increasing skills, over time, falconers' bodily and perceptual abilities were experienced as changing and as significantly shaped by the engagement with their winged companions. Falconers usually perceive the birds as active participants within these learning relationships who, like falconers, can, depending on their level of experience, act as mentors or novices.

As much as the birds play an active role in the enskilment of the falconer, they are also gradually learning skills that attune them to their human hunting companions, thus going through what Ingold, following Gibson, has termed an "education of attention", understood as a gradual enskilment of the body and with it shaping of perception and experience (Ingold 2010). Birds and humans can therefore be understood to be involved in mutually constitutive relationships of learning practices that shape the way they know, experience, and act upon the world. Through these shared tasks, a particular repertoire of skills and shared meaning emerges that entwines human and birds in what I called elsewhere a more-than-human community of practice (Schroer 2015) or in philosopher Dominique Lestel's terms hybrid community (Lestel et al. 2006).

Falconers' narratives about how they experience the cooperation with birds of prey in hunting resonate with the Uexküllian notion of subjective Umwelten. As falconer Maria pointed out to me when training her young peregrine falcon:

> What fascinates me most about the training process is that to a certain extend you need to understand the world from your bird's point of view … with time you learn to interpret the world in the way of the falcon and constantly work on your own skills to be able to anticipate her responses.

Especially when explaining the challenges involved in training and hunting with falcons and hawks, practitioners highlighted the difference between the experiential, perceptual, and sensory abilities of birds in contrast to those of humans (humans in these comparisons usually emerged as rather clumsy creatures – lacking in visual acuteness, manoeuvrability, and responsiveness) It was seen as central to have a "feeling for birds of prey", a kind of intuition and affinity for birds of prey, which was seen as an important basis for being able to learn the necessary skills. It also became evident that attention to the birds and the anticipation of their *movements* and *moods* cannot be divorced from an acute awareness of their environment as a place in which the powerful forces of the weather as well as the ever-moving air currents played a central role.

Short note on materials and methods

This chapter is based on ethnographic fieldwork in Britain conducted for my doctoral dissertation (Schroer 2015) and post-doctoral research at the Department of Anthropology in Aberdeen. My primary approach was to use longitudinal fieldwork and participant observation to actively learn from falconry practitioners about their practice. Getting hands on experience in falconry-related activities, such as various care taking, training, and hunting practices, gradually allowed me to develop an understanding of the practitioners' ways of life. Anthropologist Christina Grasseni (2007) makes a similar point, reflecting on the importance of enskilment in her ethnographic research with cattle breeders in Italy by stating, "I felt that learning to look at my host's cows was a necessary premise to accessing their worldview and sharing their practice" (pp. 208–209). The experiences and knowledge developed during this learning process were not a matter of assembling collected information as much as an acquisition of skill realised in "fields of practice" (Ingold 2010, p. 49). The material and quotes used in this article are reproduced from my fieldnotes in which I regularly recorded my observations, interactions, and conversations. They also are drawn from informal interviews and conversations I had over the years with falconers and are informed by my own encounters with birds of prey that significantly affected the way I learned to perceive and think about environmental relations. During my research, I became increasingly aware of the many ways in which birds needed to be taken seriously as active participants in shared social worlds, posing challenges to the ethical as well as methodological bases of ethnographic research. Understanding both humans and birds as participants in anthropological research, I follow and describe their relationships as they emerge through the situated (learning) practices in which they are engaged (for a discussion of methods in multispecies ethnography, see Buller 2015; Whatmore 2006). Through this approach, I do not intend to level out differences between human and avian ways of life but rather aim to contribute to a more nuanced language capable of differentiating multiple subjectivities as they emerge from shared practices.

A feeling for birds

One of the elements that practitioners highlighted as important was to have a "feeling for birds of prey" – an ability that some described as impossible to acquire solely through learning and enskilment. It was seen by many as a more obscure aspect of the practice that was difficult to articulate. Having a "feeling for birds", in short, referred to an intuitive grasp of birds, to being open towards and somehow connected to the experiential world of falcons. This feeling is, as one falconer stressed, about being able to "tune into the birds' world" and being receptive to how a bird might perceive the world in a specific moment. When asked about their experiences of discovering this affinity with raptors, a recurring narrative was that falconers discovered this "urge" or "passion" for falcons, hawks, or eagles through a key revelatory event. Falconer Katy, from Wales, for instance, described the first time she had held a falcon on her gloved fist "as a moment when something clicked". At that moment, she realised that this was what she wanted to do: "Working with a falcon and being able to make her feel comfortable around you … to make her trust you". This openness, or feeling for birds of prey, was said to be necessary to create a bond with a falcon and to establish a shared sense of communication and understanding crucial for hunting cooperation.

During my fieldwork, falconers who were said to have this intuitive resonance with birds were very often described as quite withdrawn or "awkward" in relationships with other humans. Indeed, it is commonplace in falconry literature to describe people hunting with goshawks, in particular, as socially marginal characters who tend to take on the characteristics of their feathered companions. Though not always expressed in such vivid terms, falconers have nevertheless described their relationships with birds as transformative, shaping the ways in which they move their bodies and perceive the world. As I will argue further below, a central aspect of this feeling was the ability to attune to the shifting atmospheric milieux, constituted by the forces of the weather as much as the moods of humans and bird.

Soaring in thermal currents

During my fieldwork in Italy in the hot summer month, we were sitting in the shadow of a tree sheltering from the startling heat of the midday sun. Overlooking the valley, we lethargically observed a family of wild buzzards flying high above our heads. The birds were moving on motionless outstretched wings, rising higher and higher, thereby apparently defying the gravitational pull of the earth. Circling in broad spirals, without any visible exertion, after a while, they disappeared out of sight in the bright sky above.

Alistair – a falconer I knew from my fieldwork in Britain and whom I was visiting in Italy where he was working as falconer and gamekeeper on a hunting estate – explained that the birds were catching a lift on thermal currents to reach cooler spheres in the air higher above. Thermal currents, he pointed out, were particularly strong during the hottest time of the day. To underline his point, he threw up a handful of dry leaves from the ground, that immediately spiralled upwards, following a trail of warm rising air that had escaped my awareness until then. For a few seconds, the leaves, just as much as the lines of flight of the soaring birds, made visible the transparent texture of the moving air in which we were immersed.

The air currents could be a supportive force for flying creatures in some situations as much as dangerous and difficult to manoeuvre in others. "Birds need to progressively learn to use the wind when flying", Alistair explained, "and as a falconer you need to give them the opportunity to gain as much experience as possible in varying wind and weather conditions". His gaze still on the birds above us, Alistair pointed out when a buzzard would fail to soar within the rising air and would "drop out", by being slightly thrown out of balance when exiting the air column and flying into the area of cold descending air that flows down the sides of the thermal. "When you fly your own falconry birds", Alistair warned, "you need to keep an eye out for thermals as birds love to go on a soar to cool off". Indeed, falconers often described the temptation of thermal currents for birds in the summer who when catching one would be in a state of "trance" in which they forget about the pathetically waving and despairing falconer on the ground and go up to join other birds in their spiralling ascent. The only thing left to do in such a situation is to wait until the later and cooler hours of the day when thermals fade and the birds return closer to the ground.

Thermal currents are of course not the only movements of air that become relevant when flying birds. Especially in hilly or mountainous country, the air becomes what a falconer referred to as "the white water" of the falconer. Compared to lowland country, hilly areas have a much more turbulent and irregular flow of air currents and therefore pose a greater challenge to the falconer who has to be able to anticipate how these flows influence the flights of both their own birds as well as the animals they hunt on the ground and up in the air. In fact, when talking about the air, people often drew comparisons to the element of water, or rather rivers and the ocean, possibly to give a more tactile impression of the air and its currents. Another falconer, for example, compared flying falcons to surfing waves:

> I would imagine that being a falconer looking at the air is quite similar to how the surfer looks at the ocean and its waves, not just because he enjoys their beauty but also because he can see the potentials they offer. Also just like a surfer you have to understand how the shape of the land and most importantly the prevailing weather conditions create the waves or – in my case as a falconer air currents you are looking for.

Indeed, in many ways, the air and its movements, as they progressively were revealed to me throughout my fieldwork, do share certain characteristics with the fluid medium of the river or ocean. Their movements are created by the landforms through which they are channelled and dispersed, whilst the form of the land, on the other hand, is also created through the fluid movements of wind and water. It therefore does not seem to be adequate to speak of "landscapes", "seascapes", or "airscapes" in this context as these terms seem to conjure up separate domains rather than the co-constitutive character of a world in constant formation or a "weather-world" in Tim Ingold's terms.

A crucial role in all of this plays the atmospheric phenomenon of the weather. These varying atmospheric intensities and forces have a constitutive influence on the way in which falconers and birds interact and experience their environments and other beings. Weather here, I do not understand in its scientific or meteorological sense, where it becomes turned into a quasi-thing and object of scientific inquiry. With regard to falconry practice, weather is better understood in a phenomenological perspective as directly experienced by the felt body of humans and nonhuman sentient creatures. It is such a central aspect of their immersion into the environment that it is usually not talked about divorced from context but always as embedded in the actual executions of particular tasks and in its effect on the movements of humans and birds. In fact rather than understanding the weather as something that has an effect on the world (following the logic of cause and effect), it seems to be more apt to regard it as an activity as *weathering* in which the weather *is* its ongoing effect.

Walking with falcons

For the ways in which falconers come to perceive this weathering environment not only observing birds in flight but also the direct, bodily contact during training and handling practices becomes crucial. Particularly, when carrying birds on the fist, falconers need to develop sensitivity towards the forces and intensities of the wind and position themselves accordingly in order to make the fist a comfortable perch for the birds.

On one occasion during my fieldwork in the UK at a falcon breeding station that offered internships for novice falconers, we were taken out into the hills to gain experience of carrying falcons that were being trained for the upcoming hunting season. After we embarked from the vehicle and got the birds on our fists – protected by strong leather gloves – I started to feel the wind that was blowing into the hills and it dawned on me that this was not going to be just a relaxing stroll. To me, the strong wind seemed to blow from many directions; at times, I needed to lean into it when it was blowing from the front; at others, it pushed us forcefully from the sides. The grass on the hillsides was bending and twisting in many directions and there did not seem to be a single spot within reach that offered cover from the wind.

For me, walking up the hills in this strong wind was a challenge in itself, but carrying the bird on the fist at times felt like an almost impossible task. Having a lightweight and winged creature on the fist, I felt how much an adequate positioning of my arm and fist was necessary in order to keep the bird in a steady and calm position despite the tearing winds.

The others who were walking ahead of me seemed to walk through a completely different place. Whilst their hair and clothes were blown about by the wind just as mine were, the hawks on their fists looked as though they were not moving a single feather and perched on the fist, easy and relaxed.

My bird, on the other hand, did not have a great time. When the wind was blowing from the front, she sometimes spread her wings and was shakily lifted up; at other times, the wind was blowing up her back, pushing up against her tail feathers, which threw her out of balance and made her flap her wings sharply to regain control.

After a while – probably out of concern for the well-being of the falcon – one of the others returned to help me get the bird back on the fist. Clearly amused by my clumsy inability in carrying the falcon, Jan showed me how to adjust my arm and recommended that I try to keep her close to my body, always setting her face into the wind so that the air would flow over her wings rather than get under her tail feathers. I tried it again and in the course of our walk through the hills, I managed, progressively at least, to improve the bird's position through trying always to place myself into the wind with the falcon held close to my chest, although my bird seemed never as comfortable as the others.

Carrying a falcon turned out to be a task that required skill and experience as well as the ability to be responsive to the bird whose signs of discomfort needed to be answered immediately. However, this sensory negotiation between human and bird is not only dependent on both of them but also greatly influenced by the air currents that flowed around us (and through us) and that forcefully mediated our movements.

Through considering these examples from observing soaring birds in flight or carrying a falcon close up on the fist, we see that responsive communication between falconer and bird encompasses a heightened awareness of the atmospheric qualities of the environment. These, I would argue, encompass the forces of the weather as well as the affective, emotional presence of the sentient beings involved. In the next section, I will consider an example of the initial creation of a communicative bond between human and bird, whilst focusing specifically on the atmospheric phenomenon of moods.

"A bundle of nerves": on training a goshawk

I am watching Jonathan's female goshawk from a distance in the back of his garden. Recently arrived from a breeder in Germany, the hawk is getting used to her new surroundings and is perched, unhooded for the first time, on a block perch in front of a fence that shelters the garden from the park behind it. Without the hood, and in the bright sunlight of the day, her vision is clear and acute, and her responses are fast and immediate. Her body appears tensed; she is clearly not yet relaxed, her feathers pressing against her body, both feet clutching the covering of the perch. At times, her attention is drawn to the agile swallows chasing insects above, and at times, to the restless movement of a dog exploring the bushes in the park behind her. Sensorially immersed, she is affected by the new sounds, vistas, and movements around her.

The scene I am describing is from the so-called "manning process", which in falconry refers to the initial stage in the developing relationship between falconer and bird. Jonathan, the falconer, had hesitantly invited me to join him to see his new arrival. Having had very little direct contact with humans in the breeding station, the bird was not accustomed to human company and everyday surroundings. These first weeks, in which falconer and bird were getting accustomed to each other, were very often out of bounds to me during fieldwork. This tends to be a very personal and private time in which falconer and bird are isolated from others while they gradually spend more and more time together and, eventually, with other people. Manning was described as a process of familiarisation, in which both falconer and bird learned to establish a basis for mutual understanding and communication. For falconers, it was an emotionally challenging time, accompanied by fear and the worry of failing to establish a bond with the bird. In the falconry community, goshawks are often referred to as particularly sensitive and – compared to humans – "nervous" creatures, with strong moods. They are described as "prone to panic attacks" and "shifts in personality", which can make them challenging to handle at times. Jonathan explained: "There is a certain intensity surrounding a bird that does not want to be touched or approached. A bird generally stays calm as long as you do not enter within a certain spatial distance from her"; it is crucial for the falconer to negotiate this personal space in the early process of developing a bond. As Jonathan continued: "It is almost as if when you do enter into that space you are not only touching her territory or personal space; you are also touching her".

As in the examples above, this sensitivity to the bird-in-her-environment is central for falconers aiming to establish an initial bond with a novice bird in order to anticipate her responses and to avoid negative experiences. A central aspect of this is to sense the mood of the bird that is part in creating a certain atmosphere of encounter. As Jonathan explained to me before, we went out to the garden:

> When a bird is sitting on a perch you need to be able to sense in what kind of mood she is, whether she is relaxed or intensely focussed and ready to bate at any moment. … It is a bit tricky to explain if you have not had the experience yourself … I guess it's a bit like a field of energy that surrounds a bird of prey. When you are entering the space in which she is perched you have to be sensitive to the kind of vibes she is sending you and her responds will depend on how you come across to her.

In a similar fashion, as other falconers highlighted, the need to have a "feeling for birds of prey" in order to be able to cooperate them in hunting, Jonathan is talking about the need to be sensitive and responsive to the "vibes" and "mood" of a bird.

This includes the need on the part of the falconer to closely control his or her own emotions and actions since birds are seen as:

> ...much more sensitive to what is going on in their environments ... they will sense whether you are scared or relaxed and usually respond to things so much quicker than you. Sometimes I think they have the sixth sense ... they just perceive so much more than I ever will.

Jonathan's explanation shows that falconers experience moods not so much as inner states, bound to the subjective experience of an individual, whether human or avian. Rather, I would suggest, moods were experienced as spatially tangible and palpable forces that influence human–bird interaction. Falconers often described this atmospheric character of moods through terminology such as picking up on the "vibes", "energy", or "intensity" of their encounters with birds. As described above, the birds, in turn, were said to easily pick up on a falconer's nervousness, anxiety, or anger, which had immediate consequences for the ways they would feel and respond. Part of the skill of falconry is therefore to, if necessary, be able to conceal one's true mood through controlling bodily movements accordingly. Moods, then, like the weather, are key to creating particular atmospheric milieux in which human and bird encounter each other. Let me return to my ethnography to elaborate on this.

From where I was sitting, I saw that Jonathan decided to give it a go and to start approaching the perching goshawk, hoping to be able to make her step up on his gloved fist to feed. What sounds like a straightforward task is actually a delicate and difficult aim to achieve as it requires that the bird begins to trust the falconer and to feel comfortable enough in his presence to allow such close bodily proximity. Before we went to the garden, Jonathan said he was quite nervous about this, as he had only once trained a goshawk before and that was many years ago. Yet, as he pointed out, it was important that the hawk should not sense that he was nervous, as this would immediately make her unruly too.

When commencing his approach towards the bird at the back of the garden, Jonathan was superficially putting on an air of being rather unconcerned, while always remaining acutely aware of the bird and her environment. He did not approach her directly but was watering the plants in the flowerbeds on her right, an activity that she observed from a safe distance. He did so with calm movements, avoiding any rapid or loud manoeuvres. Progressively he was working his way forward until he was quite close to the perching bird, who, whilst still not completely relaxed, seemed to tolerate him in her close surroundings.

Eventually, he slowly crouched down on his knees close to the perch, with a slightly forward-bent back he avoided towering high above her. With wings half spread out in a threatening gesture, the goshawk started to hiss at him in furious defiance. Jonathan remained calm, crouched down a bit further and ceased approaching the bird. For a while he stayed in one place, his gaze averted and avoiding any rapid movements that might startle the hawk even more. He remained in this position until the bird ceased her threatening gesture before approaching further. With his right hand, he reached out behind him, taking a piece of meat from a pouch behind his back, all the while attending to the bird in front of him, to whom he talked to in a calming voice. He then put the meat into his gloved fist and offered it to the bird, who, seemingly not yet sure what to make of this, looked at it, then at Jonathan, and back again. Jonathan lowered his arm a bit and began to talk to her, almost whispering. He encouraged her to take a bite but she refused and jumped off the bow-perch trying to bate away from him, wings beating fast and legs pulling on the jesses that were tethered to the block.[1] Jonathan outwardly remained calm – yet, as he told me later, he felt like screaming. He waited until the bird had calmed down and resumed her stand on the perch, leaving her alone for the time being.

Returning to the house, he was a bit disgruntled, not so much with the goshawk as with himself.

> I should not have tried to feed her, she was clearly not relaxed enough and I should have seen it coming that she would try to bate away from me. Now it is best to leave her in peace a bit and to wait until her mood changes. Goshawks are funny that way: sometimes they are all fury and craziness and the next minute they are cute and playful like a pet.

Aerial perspectives

Through their observations of and engagement with birds of prey, falconers learn to respond to the weather in relation to their movements and to understand the connection between landforms, air, and weather. The way falconers talk about the weather and especially the wind as well as the techniques birds and humans apply to use the wind when hunting suggest that in falconry the wind and weather take on material qualities. Air is rarely experienced as completely still, but rather in continuous movement of more or less turbulent wind currents. Amplified through the movements of the birds, air is revealed as possessing a perceptible texture, consisting of currents that at times can be supportive or dangerous and unpredictable at others. Here, earth and air are experienced as co-constituting each other, rather than as belonging to separate spheres divided by the surface of the ground.

During fieldwork, I progressively learned that to work together with an airborne creature means not to superimpose an earthbound perspective but to become sensitive to the aerial experience of the birds. The activity of the weather and its transformational forces on the movements and ways of experiencing the environment can perhaps be best described as weathering. Putting an emphasis on a weathering-world in movement, in which the weather is its ongoing effects, helps to broaden anthropological approaches to human–environment relationships, by drawing our attention to the role of atmospheric phenomena in the formation of human and nonhuman lifeworlds.

But, not only the atmospheric forces of the weather but also the spatially palpable moods of birds play a crucial role in how falconers and hawks learn to communicate. Moods were not described as emotions in the sense of inner subjective sates, detached from the material world. The term referred rather more holistically to how human or bird felt in or, was affected by, a specific situation. At once, emotionally and physically. When interacting with a bird, moods could be senses beyond the subject who experienced them as palpable atmospheric presence.

Acknowledgements

I would like to thank all of the falconers and birds who have participated in my research for their generosity in letting me part-take in their daily. This chapter reproduces material previously published in: Schroer, S. A. 2020. Fieldwork Aloft: Experiencing Weather and Air in Falconry. In Julie Laplante et al. (eds.), *Search after Method: Sensing, Moving and Imagining in Anthropological Fieldwork*. Oxford: Berghahn and Schroer; S. A. 2018. "A Feeling for Birds": Tuning into More-Than-Human Atmospheres. In S. A. Schroer and S. B. Schmitt (eds.), *Exploring Atmospheres Ethnographically*. Abingdon: Routledge; Schroer, S. A. (2019). Jakob von Uexküll (1934): The concept of umwelt and its potentials for an anthropology beyond the human. *Ethnos*, 86: 132–152. doi: 10.1080/00141844.2019.1606841. Finally, I acknowledge that my work on this chapter has been supported through my current research project, which receives funding from the European Union's Horizon 2020 research and innovation programme under the Marie Skłodowska-Curie grant agreement No. 896272.

Note

1 Jesses are thin strips of leather attached to anklets on the bird's legs used to hold the bird on the gloved fist by the falconer.

References

Azevedo, Aina, and Sara Asu Schroer. 2016. "Weathering: A graphic essay." *Vibrant: Virtual Brazilian Anthropology*, 13(2): 177–194. https://doi.org/10.1590/1809-43412016v13n2p177

Buller, H. (2015). "Animal geographies II: Methods." *Progress in Human Geography*, 39(3): 374–384. https://doi.org/10.1177/0309132514527401

Grasseni, C. (2007). "Communities of Practice and Forms of Life: Towards a Rehabilitation of Vision." In *Ways of Knowing: New Approaches in the Anthropology of Experience and Learning*, edited by M. Harris, 203–221. New York: Berghahn Books.

Ingold, Tim 2010. "Footprints through the Weather-World: Walking, Breathing, Knowing." In *Making Knowledge: Explorations of the Indissoluble Relation Between Mind, Body and Environment*, edited by Trevor H. J. Marchand, 115–132. Special issue of the Journal of the Royal Anthropological Institute. Oxford: Wiley-Blackwell.

Jefferies, Ricard 1843. "The Breeze on Beachy Head." In *Nature Abounding*, edited by E. L. Grant Watson, 139–141. London: The Scientific Book Club.

Lestel, Dominique, Florence Brunois, and Florence Gaunet. 2006. "Etho-ethnology and ethno-ethology." *Social Science Information*, 45(2): 155–177 https://doi.org/10.1177/0539018406063633

Schroer, S. A. (2019). "Jakob von Uexküll: the concept of umwelt and its potentials for an anthropology beyond the human." *Ethnos*, 86: 132–152. https://doi.org/10.1080/0014 1844.2019.1606841

Schroer, S. A. 2020. "Fieldwork Aloft: Experiencing Weather and Air in Falconry. In *Search after Method: Sensing, Moving and Imagining in Anthropological Fieldwork*, edited by Julie Laplante et al., 84–94. Oxford: Berghahn.

Schroer, Sara Asu. 2015. "'On the wing': Exploring human-bird relationships in falconry practice." Ph.D. dissertation. Aberdeen: University of Aberdeen.

Schroer, Sara Asu. 2018. "'A Feeling for Birds': Tuning into More-Than-Human Atmospheres." In *Exploring Atmospheres Ethnographically*, edited by Sara Asu Schroer and Susanne Schmitt, 76–88. London: Routledge.

von Uexküll, Jakob [1934] 2010. *A Foray Into the Worlds of Animals and Humans with a Theory of Meaning*. Minneapolis: University of Minnesota Press.

Whatmore, S. (2006). Materialist returns: Practicing cultural geography in and for a more-than-human world. *Cultural Geographies*, 13(4): 600–609. https://doi.org/10.1191/1474474006cgj377oa

6

SONIC HABITATS

Aerial nomadism and the sound of birds

Patricia Jäggi

Introduction

My multispecies fieldwork gave rise to the insight that culturally, musically, and scientifically, birdsong and other bird vocalisations that are categorised as verbal and, thus, communicative are collected, researched, learned, taught, and composed with, but that flight and locomotion sounds hardly resonate in human spheres. This chapter proposes to reflect more holistically on the sonic agency of birds. It seeks to create alternative ornithologies or eth(n)ographies, meaning less to 'highlight the significance of birds in human culture and to explore their representations' as Merle Patchett puts it (2012, 5–6) than to overcome the underlying limits and boundaries that are set by a cultural representation and appropriation of the other by a seemingly separate human culture. For that purpose, the chapter first addresses multispecies fieldwork, grounds itself in the concept of perceptual learning (Gibson 2015 [1986]), and frames listening as a mode of multispecies perceptual learning. It describes how observations in the field – which brought to the fore the diverse flight sounds of birds, such as flight calls and locomotion sounds – inspired a questioning of the ranges and filters of one's own (human) perception. In this way, the chapter outlines and explores sound as part of a bird's continuous process of inhabiting which differs from the way humans – as terrestrial beings – inhabit the world. Inhabiting the world as a migratory bird such as an Arctic tern means thinking about dwelling and habitat from an aerial and, to some extent, aquatic perspective. The chapter also takes a diversion to contemporary music pieces by Carola Bauckholt as one example of creative human engagement with these perspectives. Creating alternative ornithologies, anthropologies, or eth(n)ographies – however we label it – ultimately means to learn *with* birds, to get involved with the avian other with curiosity and openness, and to dive into alternative perceptions of the

DOI: 10.4324/9781003334767-9

one earth we share. The chapter is, thus, about multispecific sonic worlding, the formation of the world we jointly inhabit through our sensory modalities and perceptions such as through sound and listening.

Creating multispecies eth(n)ographies: fieldwork as perceptual learning

My fieldwork with wild living birds developed organically into a twofold pathway of learning (see also Ingold 2017, 23): It entailed on the one hand a *direct* learning with bird species and on the other hand an *indirect* learning as the sharing of experiences, emotions, and knowledge with humans about our avian companions. The latter encompassed semi-structured interviews as well as informal conversations, readings, listenings, and watchings of all sorts, which were all related to birds, their sounds, and the act of listening. This chapter refers strongly to the autoethnographic *direct* encounters with birds and less to the narratives of other people about their relation to wild birds.

Due to the lack of a shared verbal language – which most often makes the basis for a learning with humans – a learning with birds may be difficult to imagine. Before my fieldwork in summer 2021 in Iceland, I also would have had no idea what a learning *with* birds might mean. After the four months I spend in Iceland, I noted spontaneously back home in Switzerland:

> I realised only later that especially during my first part of fieldwork in Iceland I spent more time with (wild) birds than with humans. Through spending a lot of time with birds and having this kind of intensive sharing, of being in and inhabiting a place through listening and sound, I somehow started to kind of naturally understand sound more the way it is for birds, somehow from a non-human perspective – which is a very visual metaphor, but we do not have sonic metaphors (yet). But what I actually wanted to say is that I started to become more like a bird.
>
> *(Jäggi, 2022)*

Iceland offers vast spaces that in summer are mainly inhabited by birds. The population density of humans is small and, especially outside of human habitation, the opportunity to be in contact with birds is, in fact, greater than to encounter a human being. By spending time listening in birdscapes and listening later to the field recordings I made, I somehow, organically, started to perceive and understand the world, and the places or habitats we partially share, in a different way. This chapter thus takes its starting point in fieldwork experiences and reflects on these encounters with subpolar birds by expanding the sonic horizon of scientifically, culturally, and socially produced patterns and traditions of bird perception. Maurice Merleau-Ponty pointed to collective perceptual habituation and habits by differentiating between pure perception and the 'cultural objects' that emerge in consciousness through acquired perceptual filters and conventions. Our perceptions of the world are not free but are shaped by perceptual traditions – by a 'world born of habit, that implicit or sedimentary body of knowledge' (Merleau-Ponty 2005 [1945], 277). Our auditory perceptions of the world are likewise habituated along acquired perception patterns and filters which means that we limit our perceptual possibilities. The field journal quote above sketches the experience of an expansion of the perceptual reality which is difficult to grasp. The concept of *perceptual learning* by James J. and Eleanor J. Gibson offers a supplementary explanatory frame for the experience, whereby perceptual learning can be understood as a result of a practice and experience of a whole living organism whereby mind and body both actively engage with the surroundings (see Gibson and Gibson 1955; Gibson 1963; Gibson 2015 [1986]; Ingold 2018a, 39). As perceptual learning, the Gibsons understand the education of attention to and the discovery of the offerings of an environment which were named affordances (Gibson 1968, 138–140; Ingold 2011a, 431; Ingold 2018a, 39–40). The affordances are a shared and collective knowledge about the environment that does not know species boundaries. Similar to Merleau-Ponty, perception for them isn't something that happens in privacy but is public – or we could say collective – and isn't limited to human perception (Ingold 2018a, 39–40). In other words, perceptual learning in multispecies fieldwork means giving space to the resonances of the direct sensory experience of and immersion in an environment and in the world we share through the senses such as through listening and sounding skills.

This chapter tries to verbalise an understanding of the environment and its elements and, thus, the world not as limited to an intraspecifically shared perceptual space but as an inter- or transspecifically shared one, and this way tries to include the voices, the sonic and silent presences of animals despite the lack of a shared (human) language (Reed and Jones 1982, 412; Reed 1994, 120). In my fieldwork and multispecies encounters, listening to the living presences of animals and observing with openness and curiosity was important for being ready to share experiences and discoveries about the living world (see also Despret 2015; Buchanan 2018, 398). Field recording was not only a helpful perceptual expansion because birds in Iceland fled from me but not from an unaccompanied microphone. It was also important because making recordings has the potential to expand one's own familiar, writing- and speech-based methods with modes of non-semantic listening that focus on the vague, open, and more associative power of sound and the difficulty to grasp its meaning (for semantic listening, see Chion 1994, 25–31). Recording can therefore become a tool to amplify 'more-than-human' ways of sensorily and bodily relating to the world (Whatmore 2004, 1362). Multispecies fieldwork can be considered as a co-educational practice of multimodal ethography (without 'n') that ends up in texts and other modalities that convey the processes and moments of insight gained from learning *with* other species about the way they perceive and *live* in the world (see also Kirksey and Helmreich 2010; van Dooren and Rose 2016; Ingold 2017; Locke 2018).

Listening to a subarctic morning chorus

In May 2021, I had the chance to record the birds on Mývatn lake in the north of Iceland. At this time of the year and with Covid-19 restrictions and anxiety at work, the touristically popular place was quite abandoned. Mývatn lake is an important breeding habitat for ducks and other waterfowl in the North (Hilmarsson 2011, 301). It is characterised by its volcanic landscape, and thanks to geothermal activity, spots of the lake never fully freeze in winter. The places around the lake I visited were also humanly deserted because of the harsh wintery weather. Thanks to an early morning without north wind, I could record the birds in Höfdi, a little and unique peninsula with forest at the shore of Mývatn, which is rare in Iceland. As in the previous days, I stood out and many birds fled when I approached the lakeshore for putting my recording devices. After setting up my ambisonic microphone, I placed myself at an appropriate distance. Later, I eavesdropped on the birds through the recordings (https://soundcloud.com/user-505460012/birds-on-the-move).

At the end of May, the nights don't turn dark anymore and there is only a period of civil twilight of around four hours. It is therefore difficult to speak of a 'dawn' chorus for this northern region. Dawn chorus is an important biological and cultural phenomenon which describes the intensified bird singing in spring, which starts around 1.5 hours before sunrise and which is characteristic for the temperate zones. The dawn chorus in most parts of Europe is characterised by a natural order of songbirds that start singing, beginning with the redstarts, barn swallows, and blackbirds as early-birds and ending with the joining of starlings and species of finches (NABU 2016). To celebrate birds' worldwide early morning sounding, the first Sunday in May has become the International Dawn Chorus Day.

In contrast to the strong presence of songbirds in most of the rest of Europe, Iceland's avian fauna and, thus, the subarctic morning chorus at a bird rich area such as Mývatn is characterised by waterfowl and waders such as the Barrow's Goldeneye duck and Great Northern Diver (or Common Loon), which besides Iceland only breed in Northern America. Other European species that participated in the early morning chorus are the Grey Goose, Red-Breasted Merganser, or Eurasian Wigeon. Thanks to the forest at Höfdi, besides Redwing and Meadow Pipit, there are also songbirds such as the Icelandic Wren who can be heard in the recording excerpt (https://soundcloud.com/user-505460012/hofdi-myvatn).

The morning chorus of Icelandic birds is distinguished from the dawn choruses of Switzerland and Catalonia – the main other places we explored in our research project – by the high presence of water and wading birds, the calls of which are not thought of in a common understanding of bird chorus. We usually tend to think of birdsong of different songbirds and not of the calls of ducks or geese. Spending time listening to subpolar birdscapes put in question the dominant narrative of songbirds in our understanding of bird and sound.

Scientific and musical explorations of birds have a long tradition of focusing on songbirds. Terminology such as *birdsong* dates back to a time when bird vocalisations were turned into different types of transcriptions, either in the form of musical notes, or, more commonly, by using the ordinary alphabet for analysis. In the 19th century, bird vocalisation researchers needed to be trained as musicians to have the skills to produce notation. In the late 1920s, the condenser microphone and the tube amplifier replaced the recording horn which enabled the first field recordings of wild birds in and outside zoos. Over the course of the first half of the 20th century, audio recordings and sonograms replaced the transcription of birdsong in musical notation. In the eyes and ears of the first bioacousticians, recordings of birds had the potential to overcome the subjective and unreliable ear of the human listener. The turn from the 'musical' ear of the ornithologist to the mechanical ears of microphones in the 20th century marks the beginning of a de-aesthetisation of birdsong and a turn towards a bird's sonic or rather communicative behaviour in natural science (see Bruyninckx et al. 2012, 2018). In the 1950s, further technological development also led to a turn from the ear to the eye through the possibilities of visualisation in sonograms. Today, both bioacoustics and ornithology continue to use recording and visualisation technologies to expand scientific knowledge about the vocalisation and communication of birds, mainly songbirds. In parallel to the birdsong as aesthetic becoming a taboo topic in the 20th-century scientific development, research in zoomusicology started to reintroduce aesthetic listening, which can be exemplified by the research of Hungarian amateur ethno- and zoomusicologist Péter Szöke (1910–1994). Szöke developed tools – such as the slow-down of bird recordings – and criteria to differentiate between musical birds – such as blackbirds, skylarks, or other songbirds – and unmusical ones – such as waterfowl (Szöke 1987). Not surprisingly, Szöke was disappointed by some of the reactions to the release of his birdsong LP in 1987, which for the first time presented slowed down bird sounds as part of popular science. Funny enough, the enthusiasts of the so-called 'new music' of his age such as 'composers, aesthetes, artists' were primarily interested in the birds of his collection that he himself perceived as unmusical, such as waterbirds, gallinaceous birds, and birds of prey (Szöke 1982, partial translation Loch 2018; Szöke 1962, 50). Being confronted with the subarctic chorus of birds in fieldwork suddenly meant being confronted with auditory traditions and perceptual paradigms that in this case seem to be co-produced by an intertwinement of science and arts alike. Field recordings can enhance our listening experience, as microphones indiscriminately pick up everything that falls within their range of design. Human ears tend to filter sounds along listening habits.

By listening once again to the boreal chorus, there are other interesting sounds of birds to be explored: display flights of the Common Snipe, in which the bird stoops from high overhead and generates a pulsating, bleating sound from air passing through its fanned tail (https://ebird.org/species/comsni/L1026058); the Grey Geese's loud wing beats when flying over Mývatn; and closer sounds of locomotion on water which sound like one duck chasing away another.

Sounds of avian locomotion

I became further aware that besides calls and the many mating songs, there were avian motion sounds on the recordings. Not only could I hear them sing, call, mate but I could also hear them move on water because of a fight, of mating, playing, diving. One can hear them taking off and landing on water, hear their wings in full flight rushing over the place of the microphone.

In comparison to research about bird vocalisations, research about locomotion-induced sounds of birds is a marginal, even marginalised topic (Clark 2018; Clark 2022, personal conversation). One reason for this lack of interest is that locomotion-induced sound is, in many cases, a by-product of essentially all animal behaviour, which is why they were and often still are understood as adventitious, non-intentional, and thus, non-communicative. But there is a growing number of species found, who produce sounds by specialised morphology and/or use locomotion-induced sounds voluntarily. In this way, the sounds of locomotion can become communicative signals and then are termed sonations (Clark 2016). Specialised species that sing with their feathers are the Common Snipes, which drum with their tail and are common in Iceland (Bárðarson 1986, 269; van Casteren et al. 2010). The snipes' rather spooky sound was long associated with the call of supernatural creatures and found its way into folk tales. During the 1800s, the origin of the sound was attributed to the wings, but later, it was proven to be generated by the tail-feathers (Bárðarson 1986, 269). But it is the hummingbirds who are known as the bird family that makes the most fantastic flutter-induced, often whirring, sounds during their courtship displays (Feo and Clark 2010; Clark et al. 2011). Sonations seem less acoustically diverse than vocalisations which may be another reason why there is less interest in them. But researchers continue to find little-known examples of non-vocal sounds produced by birds in displays, such as in the Red Phalarope which also breeds in Iceland (Clark and Prum 2015, 3523). The limited scientific interest does not lie only in the quietness and non-functionality of many locomotion sounds. Clark also identifies scientific preferences and traditions that are at work: As birdsong and vocal sounds were defined as interesting and have been studied by scientists, they continue to be promoted by professors and to be studied.[1] Furthermore, the research in the field of avian sounds produced by feathers and other parts of the body of a bird and not by the syrinx resembles much more the research of entomologists than birdsong research, since insects produce all their sounds with parts of their body.

The tradition of birdsong notation for musical compositions and which is often strongly associated with the bird work of Olivier Messiaen finds its continuation and support in sonography as a helpful tool for frequency identification. A rough overview of the work of contemporary musicians who deal with birds shows a strong anchoring in the perceptual tradition of songbird song (see e.g. Doolittle et al. 2014; Taylor 2017; Rothenberg 2019). This may still lie in an aestheticising listening mode à la Szöke that categorises birds that sing virtuously and close to human tunes and rhythms as musical birds (Szöke 1962, 50; Szöke 1987), whereas other birds are excluded from a human representationalism. In the introduction to this chapter, it is stated that to create 'alternative ornithologies' would mean less to 'highlight the significance of birds in human culture and to explore their representations' (Patchett 2012, 5–6) than to try to overcome the underlying limits and boundaries that are set by a cultural representation and appropriation of the other. Had I only looked at representations of birds' sonic agency in human cultural and scientific representation, I would probably have ended up focusing on birdsong and songbirds. Birdsong seems to be a shortcut for bird's sonic agency in our cultural and scientific perceptual tradition. But it tends to exclude the more noisy, simple, monotonous, repetitive, harsh, defiant sounds, as well as a silent attentiveness of birds.

Through their music and interest in unmusical sounds of birds, the avant-garde group whose taste so disappointed zoomusicologist Szöke also worked on overcoming – or at least expanding – prevailing listening habits. A contemporary composition that continues this trajectory of expanding listening traditions can be found in Carola Bauckholt's work *Zugvögel* (2012), which is wholly dedicated to the flight of migrating birds. There is notably not a single passerine or songbird in the list of birds she used for the composition. The piece is based on the transcription of 13 birds – most of them aquatic and gallinaceous birds – who are played by five woodwinds. *Zugvögel* begins with a flying flock of swans making their circles. The birds approach and move away, creating a spatial moving effect through crescendi, diminuendi, and shifts in rhythmic patterns, as well as through beats or interferences between two performers that play only slightly different frequencies (Obermaier 2014, 48). The players also use the flap sounds of their instruments to approximate the flaps of wings. The sounds are familiar and really close but are not identical to their real-world counterparts. Through sonic mimesis and through new and unfamiliar correlations in which the sounds are woven, *Zugvögel* echoes the sound-world of birds as if in a dream state (Bauckholt 2022). Bauckholt has been working with noisier sounds of birds for many years. In other works, such as *Schlammflocke* from 2010 (Deroyer et al. 2016), Bauckholt's musical creations become like creations of new sonic habitats. Through a simultaneous proximity to and divergence from the conditioned human listening of everyday life – by amplifying sounds our historically and contextually developed perceptual filters tend to miss – her works resonate *with* the way non-humans aurally and sonically inhabit the world.

Inhabiting the world sonically

In interviews, ornithophilic people were asked to name the three birds they would choose to have with them on a deserted silent island. Songbirds that share their habitat with humans – so-called settlement dwellers such as blackbirds – were proportionally more often mentioned by these people living in Central and Western Europe than birds who inhabit territories that are not attractive for human inhabitation such as those in which seabirds, waders, and waterbirds dwell (Jäggi 2022). Sonic habituation to specific bird species that live in one's environment seems to play an important role in these preferences, and thus, in the affective relationship to birds and to one's own home. Birds that do not share habitat with humans tend to be less present in our awareness and cultural–scientific representation. Thanks to their unusual outgoing behaviour during the breeding season, the Arctic Terns and the Northern Fulmars, which both do not share their habitat with humans, make an interesting counter-example. It bears remembering that there are only a small number of songbirds in Iceland, and due to the subarctic climate, most of the birds are summer migrants. Iceland is less renowned for its biodiversity of birds than for the sometimes huge seabird colonies that breed in the cliffs or on the shores such as the Arctic Terns and Northern Fulmars. During my stay in Skagaströnd in June and July, I regularly visited the small nature park that borders this quiet fishing village in the north and got to know them better. Both seabird species breed in Iceland and are pelagic outside the breeding season, feeding on what the open ocean has to offer. Both are long-lived for birds: Arctic Terns can at least live up to 34 years (Gochfeld and Burger 1996) and Northern Fulmars up to 40 years (BTO 2022). Apart from their presence at the coastline during the breeding season, these bird species live mostly unseen and unheard to those who do not spend a lot of time on the open sea. Therefore, it was interesting to realise during fieldwork that especially the Arctic Terns are very present in the narratives of Icelanders and tourists alike.

The Arctic Terns usually arrive in Iceland by mid-May to mate and breed. They migrate annually from the Antarctic Ocean to subarctic and high Arctic breeding grounds which, despite its small size of less than 125 g, is to date known as the longest seasonal movement of any animal (Egevang et al. 2010, 2078). Funnily enough, Arctic Terns seem much bigger because of their many feathers and long wings and tail (Bárðarson 1986, 127). The tracking of Arctic Terns with geolocators has revealed that they can travel more than 80,000 km per year (Egevang et al. 2010, 2078). They breed in boreal and high Arctic and winter in the Southern Ocean such as the Weddell Sea (Ibid). No other Icelandic bird travels so far to reach its breeding grounds (Bárðarson 1986, 126–131). They usually only go ashore for breeding. During my time in Iceland, I never saw any of them walking. Their common habitat is the skies and the open sea, where they follow the areas rich in food such as krill, small fish, or plankton. Due to the fact that they make a simple nest on the ground, their eggs and youngsters are threatened by predators such as the Arctic fox, gulls, or humans. When I recorded the first Arctic Terns on 16 May 2021 (https://soundcloud.com/user-505460012/kria-budatjorn), I met them at dusk before midnight. They must have just recently landed there after their approximately 40-day northbound migration (Egevang et al. 2010, 2079). When I tried to set up my recording devices at the shore of Budatjörn, a pond in the golf course of the Seltjarnarnes peninsula west of Reykjavík, they flew up, approached me in flight and shrieked. That night, I observed that when they sit in the grass or on a stone in the water, they remain silent and all terns face the same direction, usually the direction of the wind. In contrast to their attentive silence when on the ground, Arctic Terns seem nearly constantly sonically active when in the air. One can hear this in the following recording excerpt: At the beginning, they are in the air; towards the end, they land and sit on stones on the lake (https://soundcloud.com/user-505460012/kria-budatjorn).

Arctic Terns harass any intruder, whether avian, animal, or human being. Through loud vocalisations, they try to drive the intruder out of the nesting area, their temporary habitat. They nosedive and usually attack the highest point of the intruder. The nosedives are accompanied by loud high-pitched cries followed by angry-sounding staccatos – noisedives, as it were. The word for Arctic Tern in Icelandic is Kría, which is perfectly onomatopoetic. There are even stories of Arctic Terns who used their sharp bills unsparingly and wounded intruders on the head (Bárðarson 1986, 130; Feilberg and Génsbøl 2003). When one walks through their territory, it is therefore recommended to have a walking stick and/or at least wear a hat or other head covering. Knowing about this behaviour, I first kept my bicycle helmet on when riding on a bicycle around Mývatn. But in May, they were only slightly alarmed when I showed up – they just flew over and uttered some cries to check me out. In comparison to my later encounters in June and early July – their main breeding season is in the first half of June – they feel like having been quite relaxed then. And I remember wondering about the truthfulness of these scary Kría stories.

On 5th of July, when they were actively feeding the squeakers, commuting between catching fish in the sea and the nestling on land, I walked on one of the trails crossing the breeding area that covers most of the small nature park Spákonufellshöfði. I carried a handheld microphone to record their nosedive cries. I had already experienced them in the park many times and up to this point had never felt distressed or afraid. This day they got furious about me not walking through – i.e. leaving – their space quickly enough. I used the handheld wind protected microphone as the highest point to protect myself from their nosedives and possible injuries. Finally, they attacked me not only sonically by crying at me

FIGURE 6.1 Arctic Terns nosediving – or rather noisediving – on author (Filmstills: Christoph Brünggel).

FIGURE 6.2   Arctic Terns nosediving – or rather noisediving – on author (Filmstills: Christoph Brünggel).

FIGURE 6.3   Arctic Terns nosediving – or rather noisediving – on author (Filmstills: Christoph Brünggel).

and physically by nosediving on me, but at least one of them also defecated on me. I still remember the strong smell – a mixture of old fish and acidity. I can smell it even now when I put my nose close to the non-washable fur of the microphone's windshield. After this experience, I only dared to walk on the border of their territory, a transitional zone between the nesting area and the cliffs or beach where they were much less alarmed.

I discussed this experience with Valtýr Sigurðsson, an Icelandic biologist and field ornithologist who works at Biopol, the marine research institute in Skagaströnd. For him, my observation that the later in the breeding season the more angrily the birds seemed to behave towards potential predators made total sense. The more the Arctic Terns had invested in their offspring so far, he said, the more they defend them (Sigurðsson 2021).

In Spákonufellshöfði, I also got to know the Northern Fulmars. They breed in the cliffs that enabled the place to become a nationally protected zone in 1980. The cliffs border the Arctic Terns' temporary habitat. By spending much time in this little patch of untilled land, I learned that fulmars have the complete opposite habit of uttering sounds from the Arctic Terns: when they sit in their cliff nests, they chat animatedly and loudly with each other. As soon as they are in the air, they are nearly inaudible. Their flight is silent. Often mistaken for gulls, fulmars use gliding flight and this way minimise their wing beat effort. Only on the Látrabjarg cliff which, thanks to its up to 450 m height and, thus, the distance from the rumbling sea, was I able to hear and record the soft and short hissing sound their feathers make when they fly. The sound is so tiny that it is difficult to perceive (https://soundcloud.com/user-505460012/latrarbjarg). The Northern Fulmars are, like the Arctic Terns, birds who approach with curiosity when one enters their habitat. They like to circulate along the coastline, and I also encountered them following boats on the sea. I observed them riding waves by flying directly in front of the stern waves. Through their curiosity about people, their waveriding, their bull-neck, and the tubed nostrils of their thick bill which are reminiscent of a dolphin, I started to speak about them as the dolphin-birds. In contrast to the terns, I only experienced them as curious, and they were never alarmed when I encountered them. Like the Kríur, they are known for their defence, by which they eject oil on an intruder which made their genus name: 'fulmar' comes from the Old Norse meaning 'foul-mew' or 'foul-gull' because of the foul-smelling oil. Birds have even died from Northern Fulmars' attacks, as the oil clots feathers together and cannot be removed (Bárðarson 1986, 94). It might be that climbing in their cliffs would have made them attack me. Iceland has a long and also tragic history of humans climbing the cliffs for self-sufficiency. A memorial on Drangey Island, a cliff face sprouting out of the ocean and an important breeding habitat for seabirds, commemorates people who lost their lives hunting birds or collecting their eggs.

The two bird species are in regards to their sonic agency quite contrary to one another: the terns vocalise loudly in flight and remain silent when they sit on the ground; the fulmars vocalise when they sit in their nest and do not vocalise during their flight, which is hardly hearable. Both species spend most of their time on or flying over the sea and only inhabit the land during breeding season. But what they share is that they create a temporary home or habitat in and through sound, characterised by specific spatial orders of sounding and remaining silent. And they both inhabit the earth differently than humans do. Martin Heidegger understood our being-in-the-world as dwelling: 'The way in which you are and I am, the manner in which we humans are on the earth, is [...] dwelling.' (Heidegger 2001 [1954], 147). The encounters with many birds but especially with Arctic Terns and Northern Fulmars made me consider the avian way of being-in-the-world, world-making, or worlding (Heidegger 2001; Tsing 2013; Ingold 2018b, 169). Birds inhabit other spaces and dwell in them differently than humans do – sounding and listening offers a great example to grasp this. If we think of Arctic Terns, they dwell on the earth – or rather above it – in aerial nomadism. And birds live sound and silence individually according to their own spatial understandings and sonic expressivity.

Ornithologist Robert J. Fuller states that a current understanding of habitat would not adequately include transitional zones which were important for birds (2012, 6). Habitat is usually understood as division of landscape into units or 'cover types' such as woodlands, heathland, grassland (Ibid). If we leave this underlying imagination of landscape as a divided map and turn it into three dimensions, it is the blank spaces outside the map and the sky above the landscape in a painting (Ingold 2007) that are emblematic of this huge transitional zone. From the perspective of a terrestrial mammal such as myself, the air does not seem to constitute a habitat; above all, it simply enables breathing. Fuller also writes critically with reference to Uexküll's perceptual concept of *Umwelt* from 1909 (2014):

> [D]ifferent organisms in the same location may have entirely differing perceptions (Umwelten) of the world about them, depending on their sensory apparatus, their body size, their predators, their feeding and mat-ing behaviour. The human ability to capture the essential attributes or characteristics of Umwelt is limited. Not surprisingly, few studies of avian habitats take a deeply considered bird's eye view of the environment as a starting point.
>
> *(Fuller 2012, 8)*

The flight calls of the terns and the silent gliding of fulmars inspire a different thinking about flight and sound. By switching into an avian perception of *Umwelt*, such as that of an Arctic Tern or a Northern Fulmar, the sky turns from a humanly perceived transitional zone into an avian habitat of sound and silence. If we further imagine that the air is not only the medium of birds but also the one of sounds, the sky becomes a sonic habitat. The air and the sky are a physically (through flight) and sonically (through motion sounds and vocalisations during flight) inhabited space; it is a living and a sound space. Locomotion and sounding are inseparable in the aerial dwelling as sound basically is movement, is locomotion. The Arctic Terns' breeding grounds and the cliffs in which the fulmars raise their young are a temporary sonic territory that becomes calm with the main exodus in August. By mid-September, the 24/7 breeding areas become empty and silent (Bárðarson 1986, 131; Hilmarsson 2011, 29). The Northern Fulmars stay in the open sea around the island, whereas the loyal Arctic Terns travel to Antarctica to return 9 months later to their nesting spot (Ibid).

Concluding thoughts

Listening live on the spot and re-listening to birdscapes on field recordings were defining features of the multispecies fieldwork described here. Field recording allows for an expansion of one's own listening experience through the non-discriminatory ears of technology. The above-mentioned perceptual learning through listening included amplified or expanded listening through intensified immersion in avian habitats and through audio technology. Auditory fieldwork led to the insight that when we try to understand the sound of birds, we tend to frame avian sound production as birdsong and as an alliance of symbols and communicative impulses. Perceptual traditions exclude specific species as well as so-called adventitious everyday sounds of the life of birds. Learning with birds thus meant to be confronted with the limits of perceptual traditions and trying to open up the bandwidth of my own humanly conditioned perception. By trying to understand sound *with* birds, sounds may start to move, listenings are set in motion, and the air, the sky, and heavens, this space above earth and oceans, suddenly becomes an audible space of avian sounding, locomotion, and dwelling. By listening and learning with an aerial being, not the earth but the sky suddenly becomes the place in which the listener lives, their habitat, the centre of their universe, their vantage point (following Pálsson 2020, 20). The air we breathe, the sky that turns red and orange when the sun sets, turns out to be a space that is actively inhabited. Tim Ingold recounts an excursion to the seashore with students, where they could dimly make out the forms of seabirds. They recognised the birds not as objects that moved but simply as movements that sometimes were accompanied by sounds (Ingold 2011b, 131). This shows how birds can easily shapeshift in our perceptual space and embody movement or sound in motion. Birds inhabit the air through their locomotion, and their movement is sounding – except the owls are an example of birds whose feathers are shaped in such a way as to be nearly inaudible by possible prey (Sarradj et al. 2011). Inhabiting does not only mean having a nest on the ground and sitting on a branch to sleep or in a treetop to sing. If we imagine migrating with the Arctic Terns from the subarctic or arctic northern hemisphere to Antarctica and back during a total of 130 days, the sky, the air becomes our inhabited space. It turns out to be an elusive and ephemeral habitat rather than a transition zone between points on a map or landmarks in the landscape. The sky is an inhabited part of the world together with the earth and the oceans. Birds' airiness is also sonicness. An expansion of my perceptual space by learning about sounding and listening with birds turned the world into a sonic habitat we share and dwell in. As Tim Ingold has elaborated, the relation between the listener and sound is not a relation between an agentive subject and an auditory object. Listener and sound go hand in hand in a correspondence of simultaneous and ongoing movements (Ingold 2018c). With birds, these movements can also be understood literally: understanding sound with birds supports an understanding of sound in the sense of an avian aerial nomadism, which reaches from sounding locomotion, to sound as motion, and to sound as a form of a jointly dwelling in time and space. Back on earth, this experience of sound in flight may also support a different understanding of what the sound of terrestrial bipeds might be. The work of Carola Bauckholt is one example which echoes the aerial soundworlds of birds and expands perceptual spaces that might be limited by habits and preferences such as those for 'musical' birds. The more-than-human encounters recounted here – and which are based on listening to real-world, recorded, and composed birds – hence also lead to the question of how music, art, and human sonic creativity and aural attention in general might provide a resonance space for the manifold ways in which the world is sonically and aurally inhabited. It is not only the question of how sound and listening but also how coexistence or co-inhabitation with other beings more generally might be differently understood by thinking about opening up spaces of resonance for each other.

Acknowledgements

The author would like to thank the Swiss National Science Foundation (SNSF) for supporting the research presented in this article (SNF 100016_182813/ Seeking Birdscapes: Contemporary Listening and Recording Practices in Ornithology and Environmental Sound Art).

Note

1 Personal conversation via e-mail, 28 February 2022.

References

Bárðarson, Hjálmar Rögnvaldur. *Birds of Iceland*. Translated by Julian Meldon d'Arcy. Reykjavik: H. R. Bárðarson, 1986.

Bauckholt, Carola. *Zugvögel 2011/2012: Oboe (auch Englisch Horn Rohrblatt und Jaycall), Klarinette in B (auch Jaycall), Altsaxophon in Es, Bassklarinette in B, Fagott (auch Kontraforte Rohrblatt)*. Köln: Thürmchen Verlag, 2012.

Bruyninckx, Joeri. *Listening in the Field: Recording and the Science of Birdsong*. Inside Technology. Cambridge, MA: The MIT Press, 2018.

———. "Sound Sterile: Making Scientific Field Recordings in Ornithology." In *The Oxford Handbook of Sound Studies*, edited by Trevor J. Pinch and Karin Bijsterveld, 127–50. Oxford University Press, 2012.

BTO British Trust for Ornithology. BirdFacts. Fulmar. https://www.bto.org/understanding-birds/birdfacts/fulmar, accessed on 12 August 2022.

Buchanan, Brett. "The Surprise of Field Philosophy: Philosophical Encounters with Animal Worlds." *Parallax* 24, no. 4 (2 October 2018): 392–405. https://doi.org/10.1080/13534645.2018.1546717

Chion, Michel. *Audio-Vision: Sound on Screen*. New York: Columbia University Press, 1994.

Clark, Christopher J. "Locomotion-Induced Sounds and Sonations: Mechanisms, Communication Function, and Relationship with Behavior." In *Vertebrate Sound Production and Acoustic Communication*, 83–117, 2016. https://doi.org/10.1007/978-3-319-27721-9_4

———. "Signal or Cue? Locomotion-Induced Sounds and the Evolution of Communication." In *Animal Behaviour*, 83–91, 2018.

Clark, Christopher J., Damian O. Elias, and Richard O. Prum. "Aeroelastic Flutter Produces Hummingbird Feather Songs." *Science* 333, no. 6048 (2011): 1430–33. https://doi.org/10.1126/science.1205222

Clark, Christopher J., and Richard O. Prum. "Aeroelastic Flutter of Feathers, Flight and the Evolution of Non-Vocal Communication in Birds." *Journal of Experimental Biology* 218, no. 21 (1 November 2015): 3520–27. https://doi.org/10.1242/jeb.126458

Deroyer, Jean, Enno Poppe, Marcus Creed, Brian Ferneyhough, Klaus Lang, Carola Bauckholt, George Lopez, et al. *Schlamm*. Edition musikFabrik 11. Mainz: Wergo, a division of Schott Music & Media GmbH, 2016.

Despret, Vinciane. "Why 'I Had Not Read Derrida'." In *French Thinking about Animals*, edited by Louisa Mackenzie, translated by Greta D'Amico and Stephanie Posthumus, 91–104. Michigan State University Press, 2015. http://www.jstor.org/stable/10.14321/j.ctt13x0p3s.11

Doolittle, Emily L., Bruno Gingras, Dominik M. Endres, and W. Tecumseh Fitch. "Overtone-Based Pitch Selection in Hermit Thrush Song: Unexpected Convergence with Scale Construction in Human Music." *Proceedings of the National Academy of Sciences* 111, no. 46 (18 November 2014): 16616–21. https://doi.org/10.1073/pnas.1406023111

Egevang, Carsten, Iain J. Stenhouse, Richard A. Phillips, Aevar Petersen, James W. Fox, and Janet R. D. Silk. "Tracking of Arctic Terns Sterna Paradisaea Reveals Longest Animal Migration." *Proceedings of the National Academy of Sciences* 107, no. 5 (2 February 2010): 2078–81. https://doi.org/10.1073/pnas.0909493107

Feilberg, Jon, and Benny Génsbøl. *Plants and Animals of Iceland*. Reykjavík: Mál og menning, 2003.

Feo, Teresa J., and Christopher J. Clark. "The Displays and Sonations of the Black-Chinned Hummingbird (Trochilidae: Archilochus Alexandri)." *The Auk* 127, no. 4 (1 October 2010): 787–96. https://doi.org/10.1525/auk.2010.09263

Fuller, Robert J. *Birds and Habitat: Relationships in Changing Landscapes*. Ecological Reviews. Cambridge: University Press, 2012.

Gibson, Eleanor J. "Perceptual Learning." *Annual Review of Psychology* 14, no. 1 (1 January 1963): 29–56.

Gibson, James J. *The Ecological Approach to Visual Perception*. New York: Psychology Press, 2015 [1986].

Gibson, James J., and Gibson, Eleanor J. "Perceptual Learning: Differentiation or Enrichment?" *Psychological Review* 62, no. 1 (1955): 32–41.

Gochfeld, Michael, and Joanna Burger. "Family Sternidae (Terns)." In *Handbook of Birds of the World. Vol. 3: Hoatzin to Auks*, edited by Josep del Hoyo, Andrew Elliott, and Jordi Sargatal, 624–667. Barcelona: Lynx Edicions, 1996.

Heidegger, Martin. *Poetry, Language, Thought*. Translated by Albert Hofstadter and Harper Perennial. New York: Harper Collins, 2001.

Hilmarsson, Jóhann Óli. *Isländischer Vogelführer: [Aussehen, Lebensweise, Lebensräume]*. Reykjavík: Mál og menning, 2011.

Ingold, Tim. "Against Soundscape." In *Autumn Leaves: Sound and the Environment in Artistic Practice*, edited by Angus Carlyle, 10–13. Paris: CRiSAP/Double Entendre, 2007.

———. *The Perception of the Environment: Essays on Livelihood, Dwelling and Skill*. Reissued with a new pref. London: Routledge, 2011a.

———. *Being Alive: Essays on Movement, Knowledge and Description*. Abingdon, Oxon: Routledge, 2011b.

———. "Anthropology Contra Ethnography." *HAU: Journal of Ethnographic Theory* 7, no. 1 (11 June 2017): 21–26. https://doi.org/10.14318/hau7.1.005

———. "Back to the Future with the Theory of Affordances." *HAU: Journal of Ethnographic Theory* (19 June 2018a). https://doi.org/10.1086/698358

————. "One World Anthropology." *HAU: Journal of Ethnographic Theory* (19 June 2018b). https://doi.org/10.1086/698315

————. "Comment on "Postures of Listening" by Victor A. Stoichita and Bernd Brabec de Mori." *Terrain: Anthropologie et Sciences Humaines* (2018c). https://journals.openedition. org/terrain/17547

Jäggi, Patricia. Früher die liebliche Nachtigall – heute die gemeine Amsel? *Das Bulletin. Für Alltag und Populäres.* Jan 2022. https://www.dasbulletin.ch/post/fr%C3%BCher-die-liebliche-nachtigall-heute-die-gemeine-amsel

Kirksey, S. Eben, and Stefan Helmreich. "The Emergence of Multispecies Ethnography". *Cultural Anthropology* 25, no. 4 (2010): 545–76. https://doi.org/10.1111/j.1548-1360. 2010.01069.x

Loch, Gergely. "Between Szőke's Sound Microscope and Messiaen's Organ: The Cultural Realities of Blackcap Song." *Organised Sound* 23, no. 2 (2018): 144–55. https://doi.org/ 10.1017/S135577181800002X

Locke, Piers. "Multispecies Ethnography (The International Encyclopedia of Anthropology)". *The International Encyclopedia of Anthropology*, 2018. https://doi.org/ 10.1002/9781118924396.wbiea1491

Merleau-Ponty, Maurice. *Phenomenology of Perception.* Translated by Colin Smith. International Library of Philosophy and Scientific Method. New York: Humanities Press, 2005 [1945].

Naturschutzbund Deutschland (NABU). Vogeluhr. 2016. https://www.nabu.de/tiere-und-pflanzen/voegel/vogelkunde/voegel-bestimmen/20663.html

Oberschmidt, Jürgen. "Über das Hören." In *Geräuschtöne: über die Musik von Carola Bauckholt,* edited by Jürgen Oberschmidt, 34–54. Regensburg: ConBrio, 2014.

Pálsson, Gísli. *Down to Earth: A Memoir.* Santa Barbara: Punctum Books, 2020.

Patchett, Merle. "Alternative Ornithologies." Edited by Giovanni Aloi. *Antennae. The Journal of Nature in Visual Culture* 20 (Spring 2012): 5–8.

Reed, Edward S. "The Affordances of the Animate Environment: Social Science from the Ecological Point of View." In *What Is an Animal?*, edited by Tim Ingold, 110–26. One World Archaeology. London: Routledge, 1994.

Reed, Edward S. and Rebecca Jones (eds). Reasons for realism: selected essays of James J. Gibson. Hillsdale, New Jersey: Lawrence Erlbaum. 1982.

Rothenberg, David. *Nightingales in Berlin: Searching for the Perfect Sound.* Chicago: The University of Chicago Press, 2019.

Sarradj, Ennes, Christoph Fritzsche, and Thomas Geyer. "Silent Owl Flight: Bird Flyover Noise Measurements." *AIAA Journal* 49, no. 4 (2011): 769–79. https://doi.org/10.2514/1. J050703

Szőke, Péter. *The Unknown Music of Birds (Az ismeretlen madárzene).* Budapest: Hungaroton (LPX 19347), 1987.

Taylor, Hollis. *Is Birdsong Music? Outback Encounters with an Australian Songbird. Music, Nature, Place.* Bloomington: Indiana University Press, 2017.

Tsing, Anna. "More-than-Human Sociality: A Call for Critical Description." In *Anthropology and Nature*, edited by Kirsten Hastrup. Routledge Studies in Anthropology. New York: Routledge, 2013.

Uexküll, Jakob Johann. *Umwelt und Innenwelt der Tiere.* Edited by Florian Mildenberger and Bernd Herrmann. Berlin, Heidelberg: Springer Berlin Heidelberg, 2014 [1909].

van Casteren, A., J. R. Codd, J. D. Gardiner, H. McGhie, and A. R. Ennos. "Sonation in the Male Common Snipe (Capella Gallinago Gallinago L.) Is Achieved by a Flag-like Fluttering of Their Tail Feathers and Consequent Vortex Shedding." *Journal of Experimental Biology* 213, no. 9 (1 May 2010): 1602–8. https://doi.org/10.1242/jeb.034207

van Dooren, Thom, and Deborah Bird Rose. "Lively Ethography: Storying Animist Worlds." *Environmental Humanities* 8, no. 1 (May 2016): 77–94. https://doi.org/10.1215/22011919-3527731

Whatmore, Sara. "Humanism's Excess: Some Thoughts on the "Post-Human/Ist" Agenda." *Environment and Planning A* 36, no. 8 (2004): 1360–63.

Referenced Conversations

Bauckholt, Carola. Freiburg (Germany) via Zoom Video Platform, 4 March 2022.

Sigurðsson, Valtýr. Skagaströnd (Iceland), 09 July 2021.

7

THE CHANGING GEOGRAPHIES OF HUMAN–STARLING RELATIONS IN THE SHARED SPACES OF THE ANTHROPOCENE

Andy Morris

Introduction

Despite contestations over the use of the Anthropocene as a term to describe a new geological epoch – where the mark of humans has become indelibly embedded in the very material of the earth – the incontrovertible proposition is that things have changed and changed for the worse when it comes to the fortunes of most human and non-human life on the planet. In many ways, this evokes a straightforward alarm-bell signal of an environmental threshold breached, as Donna Haraway states: 'the Anthropocene marks severe discontinuities; what comes after will not be like what came before' (Haraway 2015, 160). But the Anthropocene is also a call to recognise and reconfigure how we think about our place in all of this, including who the 'our' is as well as how they are constituting 'place'. As Jamie Lorimer has noted 'the Anthropocene describes a very different world. This world is hybrid – neither social nor natural' (Lorimer 2015, 2), and through the recognition of this hybridity, we find a world that, as Haraway describes through her theoretical refinement 'Chthulucene': '...entangles myriad temporalities and spatialities and myriad intra-active entities-in assemblages' (2015, 160). In this sense, and the key point to establish here, is that the Anthropocene can provide a context, a new vantage point from which to view the familiar and so retrace and recognise the time-spaces that make sense of the plight of these entangled entities.

In this chapter, I focus on the relations formed between two particular entities: humans and starlings, and in doing so, my aim is to at least trace the outline of a story of an entanglement of these species and the geographies they have co-produced. In doing so, the story follows co-relating humans and starlings through a 'reciprocating complexity [....] sticky with all their muddled histories' (Haraway 2008, 42) and reflects on these histories as marking out a classic tale of the Anthropocene, in

DOI: 10.4324/9781003334767-10

both the discontinuities that it marks and the insight it might offer into a new way of thinking through human–starling relations and the spaces that they co-produce and share.

The entangled times and spaces of human–starling relations

The entanglement of humans and the 'Common' or 'European' Starling (*Sturnus vulgaris*) has a history of several centuries and is rooted in the practices of hunting and eating starlings. A number of 16th- and 17th-century European depictions of this practice (Figure 7.1) exist, and it is clear that the starlings' habit of flocking in large numbers, as well as their prolific numbers per se, made them a common source of food as well as the subject of specifically developed and refined hunting methods. Whilst the eating of starlings has evidently declined over the intervening centuries, starlings-as-food still exists in the form of the French pate de sansonnet. But even through these initial tracings, the relations between humans and starlings are revealed to be more complex than simply 'hunter and hunted'; there are also more convivial and accommodating relations formed through the utilisation

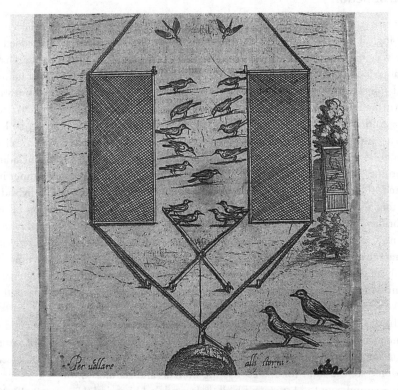

FIGURE 7.1 Caccia agli storni con la rete (Hunting for starlings with the net). Attributed to Antonio Tempesta, late 16th to early 17th century,

Source: dati.cultura.gov.it.

of starlings as, what has commonly become referred to as 'eco-systems services' providers. This role also coincides with an increase in the range of the starling as it was introduced to areas beyond Europe, primarily for the perceived benefits they might provide in insect pest control. Their introduction to sheep-rearing areas of New Zealand from 1862 saw the systematic establishment of nestboxes to accommodate the birds and facilitate their agricultural labour (Feare 1984) (Figure 7.2). This practice was also adopted in Russia where it is estimated that by 1978 there were approximately 22.5 million starling nestboxes across the country (Polyakova et al cited in Feare 1984, 279). And in 1890, 60 birds were released in Central Park, New York City, precipitating a spread of hundreds of millions of birds across the North American continent from Alaska to Mexico over the decades that followed (Feare and Craig 1999; Feare 1984).

However, the geographies of human–starling relations that I want to trace here are not simply a story of global proliferation, nor are they a story of the adaptation of starlings to human-centred environments. In thinking through the messy co-production of these geographies, the emphasis is on the negotiation and contestation of space and the role of starlings within this process. There are a number of useful reference points for this way of thinking in spatial terms, such as Haraway's (2008) development of Mary Louise Pratt's notion of 'contact zones' and the 'liveliness' of hybrid or bio-geographies explored particularly through writings on human–elephant relations (Whatmore 2002; Lorimer 2010; Barua 2017); however, the simple, central concern here is that starlings 'are active place-making agents [...] active in the production of materialised meanings around certain places'

FIGURE 7.2 Nestboxes on New Zealand Farmland (Used with permission C. Feare, credit P. Purchas)

(Bull and Holmberg 2019, 2). This emphasis on 'activity' also usefully reminds us that the relations between humans and starlings are continually unfolding, and I will strive to capture this dynamism as I outline these co-productive temporal–spatial relations from early 20th-century London to the 21st-century reedbeds of the Somerset levels and then to Rome.

The rise and fall of starlings in 20th-century London

In his 1945 work *London's Natural History*, Richard Fitter outlines the increasing presence of winter roosting starlings in London from the early years of the 20th century. Successive winters had seen not only growing numbers of birds around the centre of the city but also a distinct transition from roosting in the trees of St James Park, then to the Plane trees of various central London squares, and eventually to the buildings adjacent to these squares over the course of these winters. Writing of observations made around the time of the First World War, he states that 'this [roosting] is particularly noticeable in Trafalgar Square, where the birds gradually leave the trees round the square and join their fellows on the National Gallery and other adjacent buildings' (Fitter 1945, 130). Fitter also highlights a poetic segue from organic to human-built forms when he notes that the sculpted acanthus leaves that decorate the top of Corinthian columns were often used by roosting birds making this initial transition from tree to building (Fitter 1945). These observations are also echoed in the novel 'Antic Hay', an early work of the writer Aldous Huxley first published in 1923. Huxley, apparently quite taken by the nightly visits from roosting starlings during his time living in Paddington around 1914, reflects his own observations through the character of Mr Gumbril: '…and just at sunset, when the sky was most golden, there would be a twittering overhead, and the black innumerable flocks of starlings would come sweeping across on their way from the daily haunts to their roosting places.' (Huxley 1948, 18).

These first decades of the 20th century saw significant urban expansion in London, and the growth of starling numbers at this time has retrospectively been attributed to what is now commonly described as the 'urban heat Island' effect. The effect is created by the capacity of urban buildings to accumulate high amounts of solar energy during the day which is then retained far more efficiently at night compared with more open rural areas, thus creating a significant disparity in temperatures between urban centre and outlying areas, particularly at night-time (Wilby 2003). The 20th century also saw an increasing number of nights per year where the disparity between night-time temperatures in central London and outlying suburban areas was more than 4oc (Wilby 2003), thus making central London increasingly attractive for roosting flocks. Writing about London's increased starling numbers in the journal *Bird Study* in 1967, G.R. Potts noted that in the 20th century, central London saw '…the formation of pools of warmth and shelter that would be anomalies on the original temperature pattern' (Potts 1967, 33). In the wider spatial context, it should also be noted that the UK's starling population is significantly increased by autumn migration from populations in northern and western mainland Europe that seek the milder winter weather created by the UK's proximity to the warming effects of the Atlantic.

By the 1930s, there is a growing hostility towards starlings in London, and intrigued observation of London's winter starling roosts gives way to press coverage which reframes the presence of starlings in London through the language of disease and plague. The winter of 1937–38 sees the emergence of a spurious link between starlings and rising incidence of foot-and-mouth disease. On 26th November, 1937, *The Times* pleads for the need to 'collect evidence as to the connection between the disease and the abnormal migration of starlings from Central Europe' (*Times (UK)* 1937), and on 6 January 1938, 'a correspondent' asserts that '…action be taken against starlings [and that there are] hopes to prove whether or not starlings carry the virus of this plague' (*Times (UK)* 1938). The sentiment that starlings constitute an invading threat from overseas is only compounded in the years following the Second World War. On 25 November 1950, the *Illustrated London News* describes flocks of roosting starlings as 'an invasion from the continent [...] that darkened the sky' (*Illustrated London News* 1950). But it was a series of incidents in the summer of 1949 that became perhaps the most infamous. On 13th August, the *Times* reported that 'Big Ben was five minutes slow last night [due to] the weight of a flock of star-lings that had settled on the hands' ("Birds on the hand"). The symbolic significance of this incident was not only in its occurrence at the heart of government but in its reverberation across the UK as, at this time, the chimes of the bell were still broad-cast live to signal the start of a news broadcast. A further incident on 19th August prompted *The Evening Standard* to report the next day that 'listeners to the BBC's 9 o'clock news were startled to hear the news begin without the familiar Big ben chimes…' (Starlings Hold Big Ben Back Four Minutes 1949).

A further outbreak of foot-and-mouth disease in 1951–1952 further fuelled speculation of a link between the spread of the disease and the migration of European starlings prompting a series of debates in the House of Commons. According to Hansard records for 17 July 1952, the minister for agriculture Thomas Dugdale, under pressure to publish a report on plans to irradicate roosting starlings from London, responded by saying that 'the investigation into the starling problem is proceeding as rapidly as possible…' (Hansard HC Deb. 17 July 1952). But it was a debate on 19 February 1953 that highlighted both the desperation and ineffectiveness of the government response at this time, with Thomas Dugdale once again the focus of MPs frustrations as the following excerpt from Hansard illustrates:

> MR. DODDS: asked the Minister of Agriculture what have been the results of the experiments which have been made to reduce the number of star-lings in Trafalgar Square; what methods have been used; and how far the bird population has been reduced.
>
> SIR T. DUGDALE: Cage traps have been used in these experiments, so far, I fear, with negligible results. But final conclusions have not yet been reached.

MR. DODDS: Is the Minister aware that about 40,000 starlings find accommodation each night in Trafalgar Square; that one starling has been caught,
and that one fell in by accident? Would the right hon. Gentleman give
detailed consideration to a further inexpensive experiment, and adopt
the time-honoured method of putting salt on their tails?

SIR T. DUGDALE: I am not responsible by statute for the starlings, but I agree
that present experiments have been ineffective. It would appear that starlings are more easily trapped on their feeding grounds and not where
they go to roost.

LIEUT.-COLONEL LIPTON: Will the Minister stop fooling round with the
West End of London and concentrate on the British countryside?

(Hansard HC Deb. 19 February 1953)

In response to these debates and the growing coverage surrounding 'the starling
problem' more broadly, in August 1954, an episode of the highly popular radio
comedy programme *The Goon Show* starring comedians Peter Sellers, Harry
Secombe, and Spike Milligan was dedicated to satirising the governmental debate.[1]
The declaration that 'the inventive genius of the country was called upon, and for
three years the starlings were attacked with a series of frightening devices.' served
to set up a series of comical responses, including 'stuffed owls…wriggling rubber
snakes [and] rice pudding fired from catapults' (*The Goon Show*, 1954).

Indeed, all such measures remained largely ineffective and huge starling roosts
of tens or even hundreds of thousands continued to regularly use areas of central
London, most notably Trafalgar Square and Leicester Square. In the latter, where
cinemas and retail stimulated a valued night-time economy, daily clean-up operations and mitigation strategies such as turning over benches before dusk (Figure 7.3)
were carried out in order to try and maintain human control over these places
where the starlings, to borrow Chris Philo's phrasing, continued to 'transgress', if
not actively resist, by performing the classic role of pest species as 'matter out of
place' (Philo 1998,52). But then, during the mid-1980s, the numbers of starlings
roosting in central London fell dramatically to negligible levels in the space of a
couple of years. Whilst this decline was not the result of any of the targeted measures that had been deployed, inevitably humans were still implicated. It has been
suggested that changes to building design and the expansion of suburban London
(and therefore a growing distance between the central London heat island and daytime rural feeding grounds) could be implicated (Feare, Chris. 2018. Personal communication with author, January 25), but the principal cause appears to be a decline
in ground dwelling invertebrates and in particular the larvae of the common crane
fly (known as 'leatherjackets') which is widely acknowledged as being a key food
source for the starling (Feare & Craig, 1999; Unwin 2002; Horton 2021). As is the
case with declining insect life in the UK more broadly, a combination of the accumulated effects of post-war agricultural insecticide use and the emergent impact of
climate change are the undoubted culprits. For the leatherjackets to mature to adult
crane flies, they require soil which gets neither too warm nor too wet, making them

FIGURE 7.3 A bench mistakenly left unturned in Leicester Square, central London, early 1980s (Used with permission C. Feare).

particularly vulnerable to exactly the kind of climatic pattern that has emerged over recent decades.

The particular point I want to take from this series of events is how it sheds some light on how the Anthropocene, as a way of framing and thinking through human–starling relations, helps to make sense not only of the processes that entangle starlings and humans in this hybrid world where 'nothing remains unaltered in the event of relating' (Hinchliffe 2007, 51) but also the spaces that are produced as a result of this relating, where 'humans and [starlings] are understood to become what they are through situated and embodied interactions' (Lorimer and Srinivason 2013, 336). In this sense, these urban spaces are constituted by a continually unfolding series of actions and exchanges between humans and starlings.

Central to getting to grips with this way of thinking about starlings in the city is the capacity to move beyond the sense that animals in the city are, to return to Philo's phrase, 'matter out of place' (Philo 1998, 52) and reach for a more cosmopolitical understanding of these awkward but necessary bedfellows where they are understood as distinct entities but that 'their histories and geographies are interwoven' (Hinchliffe et al. 2005, 657). As Maria Escobar has said of pigeons in Trafalgar Square, they 'have been active participants [...] in the performative and material configuration of the square' (2013, 17). But it is also important to acknowledge that this cosmopolitical epiphany does not just play out in the city and that it is precisely in the mobilities of these constitutive encounters that we must pay our attention.

Making human–starling landscapes on the Somerset levels

Having largely written themselves out of the stories of urban London by the mid-1980s, a curious reconfiguration of human–starling relations begins to emerge at a number of locations across the UK during the early 2000s. In an interview I carried out with Wildlife television producer and author Stephen Moss in February 2018, he set the context for this emergent shift through his work on a television series called 'How to Watch Wildlife' first broadcast in March 2005 and presented by comedian and wildlife presenter Bill Oddie:

> a colleague of mine went along to an area in Gloucestershire and filmed the starlings at their roost and created this wonderful sequence of the starlings doing their antics and murmurations with Bill commenting on it [...] and that just went viral. People shared it on YouTube. People did their own versions of it on the web. One famous lager manufacturer even bought the footage from the BBC and used it in a TV advert [...] so this became something that people felt they'd missed out on if they hadn't seen.
>
> *(Moss 2018)*

This is a truncated account of events over a period of a year or so, with the Carling lager advert in question not being broadcast until the end of 2006, but it is useful to scrutinise a series of socio-technical shifts that take place in the year or so prior to the March 2005 broadcast. The first of these was the rapid rise of affordable digital cameras and video cameras alongside the growing use and quality of mobile phones with video and still image capabilities at this time. This sudden increase in accessible digital image and film making devices was also reflected in the landmark announcement by Kodak in 2004 that they were to stop production of all 35-mm film cameras, an action which saw other major camera manufacturers follow suit shortly afterwards (Davies 2004). The second shift occurs in February 2005 when the online video sharing and social media platform YouTube is launched. The success of 'How to Watch Wildlife' was also borne out by the fact that it was to become a forerunner of the highly successful BBC Springwatch series which was first broadcast just two months later in May 2005 and has continued to be broadcast on an annual basis since.

As Moss's account suggests, it is this newfound accessibility to the means of filming and then sharing films and images to a wide audience that serves to re-frame human–starling relations. More specifically, these newly emergent relations are articulated through viewing the starling roost as an affective performance, the highlight of which is the dynamic and highly synchronised aerial manoeuvres involving hundreds of thousands or even millions of starlings commonly referred to as murmurations. The function of the murmuration is thought to combine flock co-ordination and communication as well as predator evasion before the birds drop together at great speed into the roost where they gain mutual benefit from the warmth created by the confinement of collective body heat. Reimagined in a more anthropocentric, performative context however, the murmurations become the explosive finale of a show which builds slowly over the immediate pre-dusk period as starlings assemble from separate flocks of hundreds or thousands of birds converging from various directions having left their daytime feeding grounds. As the numbers swell, so the murmurations intensify both in their predator-evading dynamism and in their affective power over humans – not only through the visual spectacle but also the audible, elemental quality of the beating of hundreds of thousands of pairs of wings which, in periods of particular intensity and when the murmuration is close enough, can also provoke a visceral response as the displacement of air vibrates against the human upper body. In this moment, the starlings become situated within a new assemblage of human social gathering, filming technologies, food sharing, spectacle, and the production of collective memories. But the 'where' of these spectacles is also of particular significance. The contemporary landscapes of these human–starling interactions are now much more peripheral 'naturalcultural contact zones' (Haraway 2008) – the extensive reedbeds of wetlands, some salvaged from former peat workings, the derelict and ageing Victorian seaside piers, such as at Brighton and Aberystwyth, and the low-lying fens of eastern England, drained 400 years ago under the direction of Dutch engineer Cornelius Vermuyden. These new geographies of human–starling relations are imbued with the romantic qualities of Burke's sublime – just human enough to provide a sense of pathos but non-human enough to feel orchestrated by something bigger, wilder, and beyond.

A notable example of one of these contemporary human-starling landscapes is the RSPB's Ham Wall reserve in Somerset. The popularity of this site for starling watchers grew significantly through the 2010s as birds began to roost there regularly over the winter months, leading to the construction of additional visitor facilities, including a new fee charging car park in 2014 with more than seven times the capacity of the old 'free to use' one. This growing popularity of starling watching, along with the expanded capacity to accommodate starling 'watchers', culminated in nightly visitor numbers reaching over 1000 during the Christmas and New Year period of 2016–17 (Morris 2019). As was the case in London, once again, the starlings are reconfiguring the geographies, the prescribed spatial demarcation of the reserve coming into tension with the dominant mobility of the starlings, as site manager Steve Hughes describes:

> Yeah. I mean the starling roost here is quite interesting. It used to go to a site called Westhay Moor [...] which is a few miles away. And it would spend most of the winter there and create huge parking issues around that site. And then about eight years ago it suddenly decided to move, and it now spends most of the winter months here at Ham Wall or on our neighbouring reserve at Shapwick Heath which is run by Natural England. So, the parking issue

then moved from where it was at Westhay down to where it is now which is here. And the consequence to that we had to build bigger car parks to accommodate people. We've got tiny little rural roads and on some evenings we've had up to 1000 people here watching the starling roost.

(Hughes 2018)

The tracing of these movements outlines a lively exchange between humans and starlings as they negotiate these spaces, also adding a degree of jeopardy to the visiting crowds who are never quite sure where the starlings will roost on any given night. Hughes also relayed the disgruntlement that is sometimes directed at him and his staff from those that have turned up at the reserve only to leave disappointed when the roost has taken place on the neighbouring reserve on that night. To mitigate this, a starling telephone 'hotline' was also set up by the Avalon Marshes Landscape partnership scheme enabling paying callers access to information on where the starlings roosted the previous evening, thus providing a 'best guess' of their whereabouts.

Whilst there is not the capacity to fully engage with the associated body of work here, it is undoubtedly useful to cite the closely related notions of 'encounter value' and 'lively capital' that have been discussed in a number of accounts of human–wildlife relations (see for example Haraway 2008; Collard 2014; Lorimer 2015, Barua 2017). For Lorimer, this can be articulated through the affective power of non-human charisma and what he describes as 'epiphanies [that] are visceral and emotional but also very hard to articulate [...] a specific transformative event involving an intense encounter' (Lorimer 2015, 51). And it is the recognition of this affective, almost spiritual value within the encounter that highlights the capacity for such transformation in human–starling relations from those of 20th-century London. As Barua, drawing on the work of Collard (2014) notes, when wildlife is commodified, the nature of that being is altered: 'living organisms' dual "wild" and "commodity" lives are produced through distinct technological assemblages and spatial ecologies' (Barua 2017, 275). In this sense, it is the process of relating through these new technologies and spaces that has transformed the starlings' fortunes from pest to spectacle. But this is not simply a case of having found a spatial equilibrium of the naturalcultural, the reality is much messier than this as I now want to highlight by considering the geographies of human–starling relations in contemporary Rome.

Zoo city: the messy spaces of pest and spectacle

In many respects, the temporal and spatial trajectories of the Italian capital city of Rome are comparable to those of London. Economic growth and a rising population combined with a rapid increase in the use of motorised transport has led to a predictably similar UHI effect with a winter temperature differential between the city and proximate rural areas of 5 degrees centigrade or more (Lowen 2019). However, not only has this transition taken place more recently in Rome (during the 1950s and the 1960s), but rural feeding grounds have remained much more proximate to the city centre, thus retaining a feasible 'commuting' route for starlings between feeding and roosting sites. Whilst the effect on starling numbers is difficult to calculate, it is also useful to note that starlings were still hunted for food in nearby rural Lazio during the early 20th century, the necessity of starlings as a source of food also declining rapidly during the economic growth of the 1950s (Montemaggiori 2019). As with London, Rome is significantly more urbanised and populated than other national cities, but in addition to the exceptional UHI effect, it is likely that much of its attraction lies in its wider topographical context to migratory populations travelling from south-eastern Europe in Autumn, as Alessandro Montemaggiori explains:

> The population wintering in Rome or nearby seems to be different from the ones wintering in southern Italy, according to the dates of arrival (they arrive earlier in the season in Puglia compared to Rome). In my opinion Rome could be quite attractive because it's surrounded by good feeding grounds (olives and grape), its temperature is never very cold, there is the Tevere's [Tibor's] plain, that seems to be very important for going in and out the town, and finally the town is located near the coast where it used to be, more than 100 years ago, a very important coastal wetlands system (the natural roosting site of the species) of which today only a few vestiges remain.
>
> *(Montemaggiori 2019)*

Furthermore, despite general declines in starling populations across Europe, it has been suggested that the south-eastern European populations that swell Rome's winter numbers may be a healthy exception to this trend (Feare, Chris. 2018. Personal communication with author, November 6.). In contemporary Rome then, unlike contemporary London, winter roosting starlings have increased rather than declined in recent decades, setting the scene for a continually unfolding space where starlings have 'a very close but often conflictual coexistent relationship with humans' (Montemaggiori 2018, 41). Inevitably, this has provoked a sense of starlings' out-of-placeness in urban life, their increasing presence creating a 'flash point [...] urging their socio-spatial exclusion' (Philo 1998, 58) just as it did in mid-20th-century London.

As Montemaggiori notes above, the Tevere (or River Tibor to give it its anglicised name) appears to be a key route for starlings in navigating the city, and the river's embankment in the city centre around Trastevere and Regola became a focal point for human–starling tensions during the winter of 2015–16. The area of embankment in question is lined with large mature Plane trees on both banks and as well as providing an ideal structure for roosting, the line of trees sits within a corridor of relatively mild air, with warmer air rising from the Tevere on one side of the trees and heat radiating from the major Lungotevere embankment road and its attendant traffic on the other. As well as being densely populated by humans, this area sits within the heart of Rome's tourist centre with the Forum only a kilometre to the east and the Pantheon and Piazza Navona a similar distance to the north. By the final months of 2015, starling numbers roosting in Rome were estimated to be as high as 1.5 million, and the effects of such numbers defecating on the streets, pedestrians, cars, and scooters below forced the Rome authorities to fund contracted street cleaning company AMA to provide gangs of workers in masks and coverall suits to conduct clean-up operations each morning. Funding was also provided to try and instigate preventative measures in the form of issuing recordings of starlings' distress calls to discourage starlings from settling in the area at dusk. Initially, this was carried out by members of the Italian conservation and wildlife rehabilitation organisation Lipu with some degree of success but, as Francesca Manzia from Lipu explains, 'this intervention must be done intelligently because starlings obviously aren't stupid, and they can understand that it's false.' (Manzia 2019). Such subtle and costly measures included mounting speakers at roosting height rather than the cheaper but less convincing measure of broadcasting the calls upwards from ground level. However, starlings partially displaced by these initial efforts in the Lungotevere area relocated to the city's Verano cemetery where human sensitivities dictated that the clean-up operations were also required to follow. It soon became clear that the starlings were able to outmanoeuvre their urban companion humans. Soon, Rome's straitened public finances dictated that the use of such measures could no longer be sustained and the Lungotevere area became the starlings' once more.

In June 2016 and in light of increasing public descent over financial scandals, Rome elected a new Mayor, the populist M5S (Five Star Movement) candidate Virginia Raggi. Early hopes that this would mark a turning point in Rome's fortunes were soon dashed. A cheaper but less effective method – also not supported by conservation groups – was to use a trained Harris Hawk to scare the starlings away, but sensing this would not be effective from the start the venture was greeted with headlines such as 'Cresce lo zoo-Campidoglio: la Raggi arruola i falchi anti-storni' (The zoo-capital is growing: Raggi enlists the anti-starling hawks) (Spezzaferro 2018). This headline also speaks to a discourse emerging at this time which equated Raggi's ineffective governance with her apparent drive to utilise the more cost-effective services of animals, most notably the use of sheep to keep the city's parkland grass in check (Giuffrida 2018). Whilst the use of lasers to move the starlings was also unsuccessful, public concerns with hygiene and sanitation continued to grow as the biggest political issue in Rome, eventually leading to the resignation of the entire board of AMA in October 2019. With public protests following in response to the worsening situation, street protests mobilised under the slogan 'Roma dice basta' (Rome says 'enough!') (Giuffrida 2019) and the trending Twitter hashtag 'della colpa Raggi' (it's Raggi's fault). Ahead of elections in October 2021, headlines such as 'Rome cannot survive another Raggi term as Mayor' (Paolo 2021) signalled the end of a disastrous tenure, and Raggi was replaced by the former governmental minister for economy and finance, Roberto Gualtieri.

This brief precis of Rome's recent political landscape clearly emphasises some resonance with mid-20th-century London in that, in both places, starlings became a symbol of discontent for residents and, by extension, active agents in shaping political discourse in the city. As Paul Robbins has pointed out, through their entanglement with humans, non-humans are 'inevitably political' (Robbins 2020, 224). However, despite the possibility for many comparisons with London, there is a significant way in which contemporary Rome is situated in a temporal–spatial configuration quite distinct from that of mid-20th-century London. Because in contemporary Rome the culture of starling watching exhibited on the Somerset levels also exists; here, encounter value co-exists with clean-up costs and starlings are both pest and spectacle. Tourists now cite watching starling murmurations as one of the 'must see' features of a visit to the city with travel advice site TripAdvisor featuring posts such as: 'starling murmurations – best show in town…without doubt this will be my overriding memory of Rome' (Nik R.H. 2015), 'swirling starlings in the sunset…the most spectacular part of our stay (jroger333, 2019) and '…people just stood and watched, it was incredible' (Dyez 2020). As in Somerset, images and films taken by tourists and residents alike are commonly shared via social media sites such as YouTube,[2] and a number of internet sources provide information on the best times and places to watch murmurating starlings in Rome.

This state of situated ambivalence between humans and starlings in Rome reveals a great deal of the messiness of their spatial relations and the various contestations that play out through their formation. Just as Blue and Alexander have described the effects of Coyotes in Toronto, this cohabitation of starlings and humans in Rome has created 'a convivial if not fraught politics of coexistence' (Blue and Alexander 2017, 151). In particular, starlings in Rome express a compelling spatial agency both within and beyond the city. Once again, this can be witnessed along the Lungotevere where the starlings have not only successfully contested this space but also utilised human activity to create a socially stratified ordering of the space. Where the Lungotevere runs along the banks of the river, it intersects with a series of bridges spanning the river, thus creating a series of road junctions along the embankment. The flow of traffic at each junction is controlled by traffic lights, meaning that at each of these points there is a regular succession of static vehicles across several lanes of road, waiting for the lights to change. The heat from these static vehicles creates intense hotspots, and the starlings in the Plane trees above the road respond to this by roosting more densely at the junctions, with sparser patches of birds lower down the social hierarchy pushed to the coldest, mid-junction points. The ebb and flow of this cacophonous soundscape is clearly evident as you pass along the road between junctions creating a compelling soundscape of socio-spatial ordering. But the road and pavements below the Planes bear another powerful inscription of the spatial entanglement of starlings and humans. Having spent the day feeding in olive groves on farmland, much of which was reclaimed from their former coastal marshland roosting sites during Mussolini's 'Battle for Grain', starlings extrude faeces with a particularly high oil content, creating hazardous conditions for any ill-advised pedestrians below (Figure 7.4). The exclusion of pedestrians echoes the displacement of starlings from the marshes, reciprocated across time and space, connecting the country and the city, and the political past and present.

FIGURE 7.4 Olive stones collected from the pavement beneath Plane trees on the Lungotevere (author's own photograph).

The particular value in tracing out these diverse temporal–spatial manifestations of human–starling relations is in the richness with which they illustrate Chris Philo's observation that 'animals have inevitably been defined, categorized, interpreted, praised, criticized, hated and loved in a diversity of ways that have commonly had spatial implications' (Philo 1998, 66). However, whilst Philo goes on to argue that we might therefore 'think of a continuum between inclusion and exclusion' (Philo 1998, 66), the particular value of considering the relations forged between humans and starlings is in the co-existence of these sentiments around one species and that starlings are neither entirely included nor excluded on this basis. It is in this messier state of relations that we gain both an understanding that 'getting to know birds is not a matter of human subjects getting to know their avian objects [...] so much as a process of human-avian intra-actions' (Hinchliffe 2017, 160) but also, more broadly, that 'it is from the muddle of the middle that we have everything to learn about living' (Bingham 2006, 496). If nothing else, the Anthropocene provides valuable context here, a lens through which this process of learning can take place. In dwelling in this muddle, we see that starlings and humans, marshes and cities, olives, megaphones, political careers, and mobile phones are all implicated, woven together in a continually unfolding story. As Collard et al remind us: 'the non-human world is not a passive backdrop to human affairs and political struggles' (Collard et al. 2018, 12), it is fundamentally constitutive of them and the continually evolving geographies they produce.

Acknowledgements

I am sincerely grateful to Chris Feare, Steve Hughes, Francesca Manzia, and Stephen Moss for their valuable time and expertise. Particular thanks go to Alessandro Montemaggiori for his exceptional generosity in both regards. I would also like to thank Olga Petri and Michael Guida for all their work in setting up the 'Winged Geographies' workshop sessions from which this collection has developed.

Notes

1 'The Starlings', first broadcast 31/08/1954.
2 One of the most widely circulated images which was taken spontaneously from the back of a car in central Rome can be viewed here: https://www.inverse.com/article/41538-starlings-swarm-rome-sky-reddit-photo. Many films exist on YouTube; to search those posted by Italian filmmakers, use the search term 'Roma storni'.

References

"An Invasion That Darkened the Sky and Filled the Air with a Shrill Chorus and the Beat of Innumerable Wings: A Murmuration of Starlings at Wormegay, near King's Lynn, Norfolk," *Illustrated London News*, Volume 217, Issue no. 5823, November 25, 1950.

Barua, Maan. 2017. "Non-human Labour, Encounter Value, Spectacular Accumulation: The Geographies of a Lively Commodity." *Transactions of the Institute of British Geographers* 42, no. 2: 274–88. https://doi.org/10.1111/tran.12170

Bingham, Nick. 2006. "Bees, Butterflies, and Bacteria: Biotechnology and the Politics of Nonhuman Friendship." *Environment and Planning A: Economy and Space* 38: 483–98. https://doi.org/10.1068/a38436

"Birds on the Hand," *Times (UK)*, August 13, 1949, 4.

Blue, Gwendolyn and Alexander, Shelley. 2017. "Coyotes in the City: Gastro-Ethical Encounters in a More-than-uman World." In *Critical Animal Geographies: Politics, Intersections and Hierarchies in a Multispecies World*, edited by Kathryn Gillespie and Rosemary-Claire Collard, 149–163. London: Routledge.

Bull, Jacob and Holmberg, Tora. 2019. "Introducing Animals, Places and Lively Cartographies." In *Animal Places: Lively Cartographies of Human-Animal Relations*, edited by Jacob Bull, Tora Holmberg, and Cecilia Åsberg, 1–14. London: Routledge.

Collard, Rosemary-Claire. 2014. "Putting Animals Back Together, Taking Commodities Apart." *Annals of the Association of American Geographers* 104, no. 1: 151–165. https://doi.org/10.1080/00045608.2013.847750

Collard, Rosemary-Claire, Leila M. Harris, Nik Heynen, and Lyla Mehta. 2018. "The Antinomies of Nature and Space." *Environment and Planning E: Nature and Space* 1: 3–24. https://doi.org/10.1177/2514848618777162

Davies, Catriona. 2004. "Kodak to Stop Making 35mm Cameras." *Guardian*, January 14, 2004. https://www.theguardian.com/technology/2004/jan/14/onlinesupplement.uknews

Escobar, Maria Paula. 2013. "The Power of (Dis)placement: Pigeons and Urban Regeneration in Trafalgar Square." *Cultural Geographies* 21, no. 3: 1–15. https://doi.org/10.1177/1474474013500223

Feare, Chris and Craig, Adrian. 1999. *Starlings and Mynas*. Princeton: Princeton University Press.

Feare, Christopher. 1984. *The Starling*. Oxford: Oxford University Press.

Fitter, R.S.R. 1945. *London's Natural History*. London: Collins.

Giuffrida, Angela. 2018. "Rome Authorities Consider Using Sheep to Tackle Overgrown Parks." *Guardian*, May 17, 2018. https://www.theguardian.com/world/2018/may/17/rome-authorities-consider-using-sheep-tackle-overgrown-parks

Giuffrida, Angela. 2019. "Romans Revolt as Tourists Turn their Noses up at City's Decay." *Guardian*, April 26, 2019. https://www.theguardian.com/world/2019/apr/26/romans-revolt-as-tourists-turn-their-noses-up-at-citys-decay

Hansard. Written Answers to Questions, July 17, 1952, column 164, fifth series vol. 503.

Hansard. Commons Sitting, February 19, 1953, columns 1449–50, fifth series vol. 511.

Haraway, Donna. 2008. *When Species Meet*. Minneapolis: University of Minnesota Press.

Haraway, Donna. 2015. "Anthropocene, Capitalocene, Plantationocene, Chthulucene: Making Kin." *Environmental Humanities* 6, no. 1: 159–65. https://doi.org/10.1215/22011919-3615934

Hinchliffe, Steve, Matthew Kearnes, Monica Degen, and Sarah Whatmore. 2005. "Urban Wild Things: A Cosmopolitical Experiment." *Environment and Planning D: Society and Space* 23: 643–58. https://doi.org/10.1068/d351t

Hinchliffe, Steve. 2007. *Geographies of Nature: Societies, Environments, Ecologies*. London: Sage.

Hinchliffe, Steve. 2017. "Sensory Biopolitics: Knowing Birds and a Politics of Life." In *Humans, Animals and Biopolitics: The more-than-human Condition*, edited by Kristin Asdal, Tone Druglitrø, and Steve Hinchliffe, 152–70. London: Routledge.

Horton, Helena. 2021. "Is the UK Really Seeing a Record Daddy Long Legs Invasion?" *Guardian*, September 22, 2021. https://www.theguardian.com/environment/2021/sep/22/is-the-uk-really-seeing-a-record-daddy-long-legs-invasion

Hughes, Steve. Interview with Author, February 9, 2018.

Huxley, Aldous. *Antic Hay*. Harmondsworth: Penguin, 1948.

Lorimer, Jamie. 2010. "Elephants as Companion Species: the Lively Biogeographies of Asian Elephant Conservation in Sri Lanka." *Transactions of the Institute of British Geographers* 35, no. 4: 491–506. https://doi.org/10.1111/j.1475-5661.2010.00395.x

Lorimer, Jamie. 2015. *Wildlife in the Anthropocene: Conservation after Nature*. Minneapolis: University of Minnesota Press.

Lorimer, Jamie and Srinivason, Krithika. 2013. "Animal Geographies." In *The Wiley-Blackwell Companion to Cultural Geography*, edited by Nuala Johnson, Richard Schein, and Jamie Winders, 332–42. Oxford: Wiley-Blackwell.

Lowen, James. 2019. "Rome's Starlings." *BBC Wildlife Magazine* February 4, 2019: 62–7.

Manzia, Francesca. Interview with Author, May 2, 2019.

Massey, Doreen. 2005. *For Space*. London: Sage.

Montemaggiori, Alessandro. 2018. "Vita Segreta Degli Storni (Secret Life of Starlings)." *National Geographic Italia* 42, no.6, December 2018: 34–49.

Montemaggiori, Alessandro. Interview with Author, December 4, 2019.

Morris, Andy. 2019. "Educational Landscapes and the Environmental Entanglement of Humans and Non-humans through the Starling Murmuration." *The Geographical Journal* 185, no. 3: 303–12. https://doi.org/10.1111/geoj.12281

Moss, Stephen. 2018. Interview by Author. *Milton Keynes*. February 6, 2018.

Paolo, Michele. 2021. "Rome Cannot Survive Another Raggi Term as Mayor." *Italics Magazine*, September 16, 2021. https://italicsmag.com/2021/09/16/rome-cannot-survive-another-raggi-term-as-mayor/

Philo, Chris. 1998. "Animals, Geography, and the City: Notes on Inclusions and Exclusions." In *Animal Geographies: Place, Politics and Identity in the Nature-Culture Borderlands*, edited by Jennifer Wolch and Jody Emel, 51–71. London: Verso.

Potts, G.R. 1967. "Urban Starling Roosts in the British Isles." *Bird Study* October, 1967: 25–42. https://doi.org/10.1080/00063656709476143

Robbins, Paul. 2020. *Political Ecology: A Critical Introduction*. Oxford: Wiley Blackwell.

"Saving British Birds," *Times (UK)*, November 26, 1937, 13.

Spezzaferro, Adolfo. 2018. "Cresce lo Zoo-Campidoglio: La Raggi Arruola i Falchi Anti-storni (The Zoo-capital is Growing: Raggi Enlists Anti-starling Hawks)." *Il Primato Nazionale*, November 23, 2018. https://www.ilprimatonazionale.it/approfondimenti/cresce-lo-zoo-campidoglio-la-raggi-arruola-i-falchi-anti-storni-97193/

"Starlings Hold Big Ben Back Four Minutes," *London Evening Standard*, August 20, 1949.

"Suspects Among Birds," *Times (UK)*, January 6, 1938, 15.

Unwin, Brian. 2002. "Britain Proves too Hot and too Wet for Vanishing Crane Fly." *The Independent*, August 19, 2002. https://www.independent.co.uk/climate-change/news/britain-proves-too-hot-and-too-wet-for-vanishing-crane-fly-173887.html

Whatmore, Sarah. 2002. *Hybrid Geographies: Natures, Cultures, Spaces*. London: Sage.

Wilby, Robert. 2003. "Past and Projected Trends in London's Urban Heat Island." *Weather* 58, July, 2003: 251–59.

8

THE PUBLIC LIVES OF PIGEON PASSENGERS

How pigeons and humans share space on a train

Shawn Bodden

Just your ordinary, workaday pigeon

A group of friends wait for the train. Around them, the station is quiet. The friends chat, laugh, wait, scan their phones, look around, wait some more. One pauses, nudges another, then points to a pigeon standing right at the edge of the platform, well past the yellow safety strip. They hear the rumble of the inbound train, but the pigeon stays put. One of the friends takes out their phone and starts a video as the train approaches the station—as the train approaches the pigeon. 'Oooh-oohoohooh!', they exclaim as the train rockets just past the tip of the pigeon's beak.

The pigeon is unharmed and unfazed. In fact, it turns and quickly approaches the doors of the train. The group of friends follow, exclaiming as they record. The pigeon waits alongside a man queuing at the doors: when they slide open, the pigeon steps up to the edge and boards the train with a measured hop. The friends laugh and exclaim again. This is something unexpected. The pigeon, it seems, is taking the train.

The friends board the train after the pigeon, chatting animatedly and curious to see what happens next. The pigeon turns right, walks to a small divot where two of the cars connect, then hops in. Laughter and shrieks of amusement burst from the friends. They go to find their own seats and to post their video of this 'workaday pigeon' online,[1] where it's captioned with a French idiom, 'Métro, boulot, impôt, dodo': *commute, work, taxes, sleep*. Like them, this pigeon is on the daily grind. The video is shared and re-blogged—if not exactly widely, at least enthusiastically—and online commentators share in the friends' amusement.

'It is so deliberate in finding that specific hole on the floor, as though it does this every day and has a favorite "seat" on this one particular train', one commentator observes, to which another replies wryly, 'Pretty much like all LA metro riders'.[2] Others affirm this same basic observation—that this pigeon sure seems to

DOI: 10.4324/9781003334767-11

know how to board a train—while others make jokes about the rules it ignores: fare-dodging; not letting other passengers off before boarding; and, indeed, not flying. 'But why?? you can fly, dove...', one person wonders for the rest of us.

Although apparently a surprising sight, videos of pigeons riding trains like this one have been shared from cities worldwide—Santiago and Stockholm, New York and Tokyo. Alongside the comments expressing shock at the novelty of a pigeon in the apparently human-only space of a train car, others share experiences from their own uneventful multispecies commutes. In fact, in his accounts of the 'public life of the street pigeon', Eric Simms recalls travelling with a pigeon on the Tube in London as early as 1965:

> At Kilburn station, which is above ground, as the doors opened a street pigeon walked into the coach where I was sitting. It watched the doors close and remained calm and without moving on the floor. It ignored the next station but, as the doors opened at Finchley Road, much to the amusement of my fellow travellers and myself, it flew out on to the platform quite unconcerned and joined up with three or four others. Commuting appears to be no human prerogative!
>
> *(1979, 14)*

In 1995, a series of letters to the editors of the *New Scientist* further confirmed that train-riding pigeons are not exactly new: 'In my experience, the sight of pigeons hitching a lift on the underground is nothing unusual. I too have often travelled from Paddington ... not infrequently in the company of a pigeon, sometimes even two' (Howlett et al. 1995: online). These pigeons are described as 'nonchalant' and 'calm as you please' while waiting at the door, ignoring crisps proffered by excited tourists, and alighting 'with purpose'. One letter recalls a specific pigeon 'of light reddish colouring' prone to boarding at Paddington and routinely disembarking at the following station.

In these accounts, it is not just the presence of pigeons on trains but also the details of their behaviour that catch human passengers' attention. These pigeons are not, for instance, seen to be looking for food—a notable observation given the centrality of food and food *waste* in human interactions with, and understandings of, pigeons. Practices of pigeon-feeding feature prominently not only in representations of urban pigeons as convivial companions but also in campaigns to 'cleanse' public spaces of pigeons as dirty 'rats with wings' (Blechman 2013; Jerolmack 2013; Escobar 2014; see also Howell in this volume). As a recent popular-science account of train-riding pigeons puts it, 'Pigeons hang out wherever people drop food ... To a pigeon, a subway car is a warm box full of sleepy humans and spilled food' (Mosco 2021, 149–150).

The imaginative geographies of humans about pigeons—the ways we understand their activities, what they do, and where they go—appear distinctly food- and waste-centric (Philo and Wilbert 2000; Jerolmack 2008, 78). As Colin Jerolmack

(2013, 70–75; drawing on Douglas 1984) argues in his sociology of pigeon–human encounters, humans impose their own normative spatial expectations on nonhuman animals, and their failure to conform to expectation can lead to persecution as 'matter out of place': pigeons, he observes,

> have become particularly despised urban trespassers partly because they, in all their animality, are so *public* … they do not even retreat to sewers, trees, or parks to defecate, mate, and live, as do so many other animals. Further, these birds may evoke discomfort or even nausea by scavenging humanity's refuse.

Yet, the train-riding pigeon from the video introduced above does not appear to become a problem—at very least, not yet and not for the people sharing the train with it—and the accounts of train-riding pigeons that get shared only rarely echo the disgust, worry, or opprobrium we might expect about such 'out-of-place' pigeons. Instead, these stories display a sort of ambivalent curiosity and amusement about the out-of-place pigeons who seem disinterested in 'pigeonly' activities like eating, rummaging, or—one might expect—flying on a train. In fact, these stories tend to emphasise just how *little* happens: 'Neither us passengers, nor the birds, took the slightest overt interest in each other and the pigeons always hopped off at their stop rather than winging it' (Howlett et al. 1995: online). In fact, in the video above, it seems fair to say that the pigeon may be out of place, but it is exactly how it *finds* a place that interests and entertains the groups of friends so greatly.

In the ambivalent encounters between pigeons and humans on trains, the public lives of urban pigeons are not straightforwardly 'problematic' or 'convivial': instead, such encounters constitute complex sites of attention, reaction, and reasoning about the activities of other human and pigeon passengers. Although easy to dismiss or overlook, especially in comparison to the dramatic controversies and conflicts that tend to dominate theoretical reflections on the place of pigeons in society (see Jerolmack 2008; Amir 2013; Blechman 2013), staying with such mundane encounters can offer insight into the interactive and everyday practices of sharing multispecies space in everyday life. In this chapter, I develop an alternative approach to understanding the public life of urban pigeons in terms of an open-ended process of 'public reasoning' (Barnett 2017) about how to live with and alongside other species. Such a shift not only keeps sight of the curiosity and ambiguity that characterise everyday life in the multispecies city but also works to push understandings of 'imaginative geographies' away from an image of determinative 'underlying' social frameworks (cf. Jerolmack 2008, 89) towards their recognition as ongoing and interactive processes of sense-making—in which pigeons, too, take part.

In the following sections, I first situate the claim that urban animals like pigeons are treated as 'out of place' in 'human space' within a broad literature on relational ontology and theories of entanglement, before turning to recent calls to consider the 'limits of relationality' (Rose et al. 2021, 13) and practices of 'ethical' distancing (Ginn 2014; Giraud 2019). I then argue for a shift from ontological and epistemological debates about human and more-than-human society to a prax-eological analysis of spatial logics: how people—and pigeons—reason with and about space in everyday life. Drawing on ethnomethodological (EM) studies of interaction as 'practical sociological reasoning' by everyday members of society (Garfinkel 1967, 11–12), I explore the implications and challenges of situations in which humans and pigeons reason together—sometimes fraught, sometimes friendly occasions of multispecies membership. By applying this approach to think through videos of human–pigeon interactions on trains, I argue for a situated and open-ended understanding of pigeons' winged geographies, where pigeons learn how to use wings in a human world, but humans learn how to live in a winged world as well.

Finding a place on the train

The responses of laughter, pointing, curious questions, and amused stories shared by human passengers about train-riding pigeons clearly mark them as 'out of place', unexpected if not necessarily uncommon. For Jerolmack (2008, 88–89; drawing on Griffiths, Poulter and Sibley 2000), this is indicative of a more general status of pigeons as urban transgressors, who 'defy our conceptual categories and attempts to situate them in certain spaces'. In an important move, Jerolmack (see also Philo 1995) links the perception of pigeons as 'trash animals' with anthropologist Mary Douglas's (1984) influential account of the social construction of 'dirt' as 'matter out of place' within human social schema. For Douglas (1984, 36) and the many schol-ars who have adapted her work (see Ablitt and Smith 2019, 868–869 for a useful and critical review), studying perceptions and treatment of 'dirt' offers a method for uncovering social taxonomies that govern moral norms and behaviour: she writes, 'Dirt then, is never a unique, isolated event. Where there is dirt, there is system. Dirt is the by-product of a systematic ordering and classification of matter, in so far as ordering involves rejecting inappropriate elements'. In this way, perceptions of pigeons as 'pests' or 'trash animals' become indicative of 'the underlying imagi-native geography of the modernist constitution ... [which] places firm boundaries between nature and culture and views transgressions of these boundaries by animals as pollution' (Jerolmack 2008, 89–90). Pigeons as 'trash animals' represent and cre-ate 'social disorder'.

In particular, the 'underlying imaginative geography' that pigeons are taken to contravene consists of a set of familiar distinctions between 'nature' and 'culture', 'public' and 'private', 'human' and 'nonhuman' popularised by Bruno Latour (1993, 10–11, 2013) and other influential Science and Technology Studies scholars like Donna Haraway (1991, 2016) and John Law (2004) as indicative of the 'Modern' worldview, one in which the world is organised according to 'two entirely distinct ontological zones: that of human beings on the one hand; that of nonhumans on the other'. These scholars and others influenced by their work have developed a range of alternative ontological perspectives that emphasise relation, hybridity, and entanglement in an effort to acknowledge the creative agency of nonhumans and foster more ethical forms of living (Whatmore 2002, 165–167; Bennett 2010). While there is considerable variety in these accounts, they share a broad preference to theorise the 'problem' of animals like pigeons in an ontological mode and, as a consequence, recommend alternative, more ethical modes of living with nonhumans that turn on recognition of the world as ontologically, that is, 'really' entangled (see, for example, Law and Singleton 2013, 490). Thus, humans are encouraged to learn 'to appreciate the "contaminated" biodiversity of our hybrid landscapes … to model more sustainable human and nonhuman cohabitation' in order to confront and reject 'hackneyed' imaginative geographies that insist on division between human and nonhuman (Jerolmack 2013, 237).

In recent years, however, a number of critiques have been made about the ethical assumptions behind such prescriptions of relationality. As cultural theorist Eva Giraud (2019, 7; Hollin et al. 2017) observes, 'acknowledging that human and more-than-human worlds are entangled is not enough in itself to respond to problems born of anthropogenic activity' and can in fact inhibit intervention by resisting ethical or political claims seen to lack 'complexity'. Giraud's work offers a number of important observations about the tensions that arise between theories of ethical relation and activist practices to respond to concrete, situated problems. She draws attention to the creative and constitutive role of practices of *exclusion* in negotiating ethical ways of living alongside other, sometimes dangerous animals (Ginn 2014; Giraud 2019, 15–20), while also noting how relational affects can be exploited to do violence to animals as well, such as in the exploitation of laboratory beagles (Giraud and Hollin 2016). Giraud's critique (2021, 42–43, 2019, 18; see also Lynch 2019) shifts the register of questions about ethical relation, exclusion, and obligation to consider the tactics and strategies people use to *respond* to situated encounters and to remind theorists that it is often 'those already facing poverty and discrimination who face the burden of negotiating … experiments in multispecies living' in practice.

Giraud's work resonates with a recent move to call into question the 'new positivism' and 'affirmationism' of more-than-human, vitalist, affective, and 'additive' accounts of human entanglement with the nonhuman in cultural geography (Philo 2017, 257–258, 2021, 77). Proponents of 'negative geographies' (Rose et al. 2021, 3; Philo 2021) have argued for serious consideration of the challenge of limits and distance, 'the question of what can be done' and how 'the very question of doing [is] situated within a context of the doable'. In this view, relation and entanglement serve less as an ontological corrective to flawed social norms, and more as a situated and ongoing *challenge* to which humans and nonhumans respond (see Massey 2005, 179; Bodden 2022).

This observation recommends an attention to the experience of 'conjunctive and disjunctive relations' in the ongoing and practical responses to *problems* of plurality and entanglement (Savransky 2021, 152–153). Even more crucially, it challenges the assumption within ontological modes of theorising that mapping 'complexity' or delimiting 'general metaphysical properties' can lead straightforwardly to more just or ethical forms of life (Barnett 2008, 187). Such theoretical

models offer minimal insight into the situated, practical, and plural 'forms of public reasoning and assessment' through which people negotiate and justify particular
responses to 'entangled relations' or nonhuman 'transgressions' in practice (Barnett
2017, 271).

Bringing this to bear on analyses of pigeons' place—or 'out-of-place' status—
in human space, the assumption that the discourses and beliefs of 'claims-makers'
(Jerolmack 2008) proceed from underlying cultural and spatial norms ignores and
obscures the active and lively role of claims-making within situated processes of
public reasoning by treating claims as mere evidence of mistaken, 'hackneyed'
ontologies. In contrast, an 'action-oriented' perspective that takes 'talk-data not as
a mirror on the mind, but as a resource for understanding the ways in which ordinary reflective, deliberative capacities of subjects-in-practice fold together habit and
reflection' insists on understanding what people are doing in making, responding
to, and 'processing' claims in and as action (Barnett et al. 2011, 116; Barnett 2017,
253). The public life of urban pigeons not only involves diverse and shifting relationships with humans, including campaigns to exclude 'rats with wings' but also
debates over 'humane' methods of pigeon population control, online communities
coordinating pigeon-rescues, dedicated pigeon-feeders, and exasperated critics of
pigeon *over*-feeders seen to be hurting the birds by attracting negative attention to
them (Blechman 2013, 242; Jerolmack 2013). As forms of impassioned, embodied,
and interactive claims-making, these practices make a pigeon's place in human society into an open-ended, recurrent, and negotiated question—one in which pigeons
might, in their own way, have a say.

The troubles that arise through ontological prescriptivism elide not only the
public reasoning of humans but also the role nonhumans and their practices play in
this process. In her analysis of the complicated and ambivalent 'affective landscape'
surrounding urban colonies of kittiwakes along the Tyne river, Helen Wilson (2021,
5–7; drawing on van Dooren 2019) suggests seeing bird practices as 'avian claims
to space'. As she observes, the conflicts that emerge over urban birds are less often
between 'human and bird' as between 'different groups and positions, competing
forms of legislation and protection, and different understandings of what constitutes
appropriate action, not only between those that disagreed on whether kittiwakes
have a claim to urban space, but also between groups that might otherwise be considered allies in their concern for recognising avian claims' (Wilson 2021, 15). The
question of how to live with urban birds turns, in part, on the 'processing of' and
reasoning about avian activities as claims to *certain* forms of life and relationship with
humans: claims to community 'without prior guarantee of felicitous outcomes …
a claim to community that might well be queried, generate dispute, or simply fail'
(Barnett 2017, 52, 61).

The ways people make sense of and respond to avian activities in situated
encounters with urban birds (Wilson 2011, 2017, 464)—and, in return, how birds
make sense of human activities—should not be construed as an epiphenomenal
by-product of social schema but rather recognised as an important part of the contingent and contested negotiation of multispecies community as a challenge, the

ongoing and 'ever-contested' terms and politics of sharing space (Massey 2005, 142). The ambivalent and uncertain encounters that emerge between pigeons and humans on trains are thus useful sites from which to reflect on the 'public expression and exchanges between persons or between nonhuman animals' that drive the *ethological* processes of 'determination and understanding of human and nonhuman "forms of life"' (Leys 2017, 17–18). To understand how a pigeon's 'place' in society is negotiated day-to-day as a live concern, greater attention should be given to the embodied and interactive 'practical sociological reasoning' (Garfinkel 1967, 11–12) through which humans *and* pigeons work out how to live together by responding to one another. In the following sections, I recommend ethnomethodological studies of interaction as one way to think about and inform these debates through the analysis of actual encounters between humans and pigeons on trains—the situated and ongoing negotiation of the practical problem of multispecies membership through practices of relation and exclusion, inquiry, and response.

The practical problems of pigeon passengers

Ethnomethodology (EM) is a school of thought that takes social order to be an ongoing, practical, and joint accomplishment by members of society (Garfinkel 1967, 2002). Although its general approach resembles theories of 'social-constructedness' and 'performativity', EM insists on the local contingency of the social systems, contracts, or rules that other theoretical approaches grant conceptual priority: even where such rules appear to be a 'standardized feature of a normal environment', their situated relevance, consequence, and meaning are contingent on and shaped by 'specific courses of interaction' (Lynch 2012, 228). Rules and norms are *resources* for reasoning in and about a situation, rather than determinative of it, and treating them as 'a stable field of normative possibilities for actions in a given situation' proceeds by treating people as 'cultural dopes' incapable of assembling and acting on their own judgements (Lynch 2012, 228; Garfinkel 1967, 66–69). 'Culture' in EM is not taken as the explanation for action but rather as a 'topic' of inquiry (Francis and Hester 2017). Thus, EM directs analysts' attention to the situated details of interaction to understand the 'methods' used by members of society to understand, interpret, categorise, and organise their activities and the events they find themselves involved in.

Although EM is itself emphatically not a method—it is, rather, the study of members' everyday methods—the sequential analysis of interaction recorded in video has proven a popular and distinctive ethnomethodological technique to draw attention to the embodied and multimodal nature of social interaction and public reasoning (see Laurier 2014, 2016). By studying communicative action, including but crucially *not limited* to human language, this approach has also been used to examine social interaction between nonhuman animals, highlighting the importance of mutual orientation, responsiveness to movement, and embodied accountabilities of action in nonhuman communication (Mondada 2018, 101–103). Despite a tendency to prioritise human interaction and membership, EM studies of interactions between humans and dogs—at play (Goode 2007), going for walks (Laurier et al. 2006, 21), and in their work as trained assistance dogs (Arathoon 2022)—make clear that 'interspecies coordination' is a routine part of everyday life. Social 'membership' is not contingent on correctly understanding the 'internal state' of another or sharing a catalogue of 'cultural norms' but rather consists of mutual reaction and interpretation *subject to revision*: processes of reasoning and learning together (Goode 2007, 14; Francis and Hester 2017). Multispecies interactions constitute forms of embodied and public reasoning through which 'humans and other animals are alive to aspects of the world' including each other (Leys 2017, 132).

In this case, whether and *how* a pigeon is 'out of place' in a train is to be discovered locally by those 'party to the scene', not pre-determined by a cultural norm. This helps to understand and recognise the diverse ways of responding to and reasoning about train-riding pigeons shared in the video, online comments, and letters discussed previously. In some situations, pigeon passengers are 'out of place' in a banal way, meriting little more than an eyebrow-raise. In others, pigeons like the regular commuter 'of light reddish colouring' gain a place—between Paddington and the following stop—by dent of their observed habits. In the video, another pigeon recognisably 'finds a place' as marked by the friends' laughter and our own as viewers when greeted with certain practices of waiting, boarding, and 'taking a seat'. These techniques offer no guarantee of success, but they play a part in how this pigeon becomes a humorous and surprising presence on the train, rather than a problem to be dealt with for the other, human passengers. Moreover, the group of humans observing the pigeon deploy a range of techniques themselves to accomplish the situation as such: their laughter—carefully timed with the pigeons' practices of 'finding a place'—is itself a public evaluation of the situation as a humorous one, and their measured distance from the pigeon enables its continued autonomy and ease.

In this case, 'keeping distance' plays an important role in allowing the pigeon to become an 'unusual passenger' rather than, for example, a 'dirty pest' to be expelled. To appreciate this, we might imagine how the situation may have played out differently had the pigeon begun walking about people's feet 'begging for food' as they're seen to do in parks (Jerolmack 2013, 27–29). For some particular 'party to the scene', this could become a 'problem' or an 'intrusion of personal space' within the normatively organised space of a train car, where each autonomous passenger has their own 'place'. However, it is important not to lose sight of the fact that the normative order of public space is established in and as situated interaction (Jayyusi 1991, 246), often through subtle gestures and embodied responses to the co-presence of strangers (Laurier and Philo 2006). It may be suggested but cannot be fixed a priori by cultural norms or the physical design of the train car.

As Muñoz (2021a, 610–611, 2021b, 9–10) observes in his study of experiences and embodied practices of disabled passengers on public transport, both apparently 'positive' practices like recognition and 'negative' practices like avoidance play an important part in enabling others to travel as passengers: 'supportive interchanges of involvement' are underpinned by a maintenance of distance from those 'acknowledged as an autonomous fellow passenger'. As in the video, stories shared about pigeon passengers tend to emphasise how *little* happens and how disinterested the pigeons seem in either people or food: taken as 'instructed action' (Garfinkel 2002, 105) about how to distinguish a pigeon 'passenger' from a 'dirty pest' or a 'trapped animal', such accounts provide insight into the practical problems humans and pigeons might encounter while travelling together and the techniques of *attentive avoidance* used to resolve them. This avoidance is not a matter of course or a reflexive aversion to a 'dirty pest' (cf. Jerolmack 2008, 88) but a reasoned response to a pigeon's presence as a 'claim to space' and its own place on a train (Wilson 2021)—a claim formulated and interpreted through its situated practices and embodied disposition.

This encounter with a pigeon passenger recommends a more nuanced—and perhaps more tentative—story about the 'place' of pigeons and other urban animals in society than those offered by theoretical accounts that insist on either ontological relationality or division. Nonetheless, it is important to emphasise that the video is not intended as an exemplar 'case' of how pigeons and humans encounter one another on trains. It shows, more simply, that and how such encounters involve open-ended negotiation which might productively be likened to the processing of and response to claims. This point can be extended, however, by considering a contrasting encounter in which a different pigeon passenger is found to be problematically 'out of place'—and removed from the train.

Travel in winged worlds

A pigeon has gotten in the train, and a man is trying to catch it. It's not clear how the altercation began, but now, the pigeon eludes the mans' hands by walking across a row of seats and then—when the man reaches closer, hand cupped to catch the bird's back—taking flight. The pigeon lithely manoeuvres between the man and a pole, flies higher to sail above the head of another passenger—who ducks slightly and covers her face—before colliding with a vent in the ceiling. It flies on, through an open area in the adjacent car. Two passengers watch the pigeon fly past as they take their seats; another grabs her bag, stands up, and retreats briskly to the other car, away from the pigeon. She smiles sheepishly at the camera as she passes and then looks back to check on the bird's whereabouts.

Some of the other passengers have stood up now and watch the scene, while a few withdraw further down the car. The pigeon lands on the floor, and a man goes to grab it: he follows it with his outstretched arm first as it toddles around his legs, then when it flies up and into a window, and still when it flees under the nearby seats and a row of sitting passengers. As the pigeon and its pursuer work their way down the car, two more men join the chase, grinning. Other passengers step aside, raise their legs, and vacate their seats to make room—to get out of the way. Finally, the man manages to shoo the pigeon out from under the seats, and another catches the bird against the ground in one hand. He's on his mobile, however, so he turns and hands the phone to another passenger before reaching down with his other hand to scoop up the restrained bird, who wriggles and rustles its wings against the floor. The man arranges his hands to hold each wing tucked against the pigeon's back and then stands and turns towards the nearby door. His timing is impeccable: not 10 seconds later, the train pulls into the next station, and the doors slide open. He retracts his right hand from the pigeon and reaches his left arm forward and up: the pigeon takes flight.[3]

Unlike the previous video, this 'out-of-place' pigeon *is* removed, and a number of human passengers devote considerable effort to this end. However, when the video is uploaded to the internet, the poster does not refer to the expulsion of a dirty 'trash animal' or 'pest' but rather describes the scene as one in which a trapped bird is 'rescued by passengers'. This description of the scene is made possible through a range of observable details through which the passengers 'party to the scene' make sense of, interpret, and respond to the encounter. The man who catches the pigeon, for example, handles the bird carefully: he restrains the bird against the floor but gingerly enough that it can still move about in his hand. By removing one hand and raising the other, he fluidly 'releases' the pigeon, supporting its flight away. The bird's removal is accomplished as a 'rescue' through public recognition of its distress—its fluttering into the ceiling and windows—and response by handling it with measured care. No one, for instance, tries to swat, harm, or kill this pigeon.

Indeed, despite the common claim that *flight* in particular marks birds out as categorically 'liminal', 'disorderly', or 'at odds' with human ways of life and 'urban order' (Wilson 2021, 8; van Dooren 2014; Jerolmack 2008, 89), here the pigeon's flight becomes an important resource for making sense of and responding to the situation—for seeing it as a bird in need of help, in contrast to those pigeon passengers who stand to the side apparently wanting nothing of the humans nearby. Claims about 'moral', 'social', and 'cultural' transgressions by birds on the wing may arise, but they too must be treated *as claims*: situated, passionate, and public 'doings' used to call for and justify certain courses of action and modes of response (Barnett 2018, 7). They are methods of public reasoning about how to respond to the challenge of multispecies membership when and as it arises. The frequent interpretations of such actions as necessarily manifestations of an underlying 'collective desire' to exclude (Jerolmack 2013, 227) or even a 'war against the pigeon' (Howell, this volume) make it too easy to ignore or dismiss the criteria and conditions through which a given response is developed and, indeed, critiqued or revised during the practical work of learning how to live with pigeons.

Although a story of someone grappling and ejecting a pigeon from a train may, taken out of context, be seen as human violence and power over a nonhuman transgressor, these actions in this context become practices of 'how to help' through the embodied reasoning and attentive response of the passengers (recalling Ginn 2014). Rather than insisting that these actions are beyond critique, this is merely to argue that evaluations, criticisms, and suggestions of alternative courses of action to live alongside pigeons better should not target abstract moral categories rooted in various 'ontologies' of the social but rather offer critique in an 'ordinary' register

(Barnett 2014, 157) that seeks to inform and justify alternative courses of action next time, for example, someone finds themselves sat beside a pigeon on the train.

Shared and enacted in this way, even exclusion and separation can serve as 'practices of more-than-human cohabitation' and vernacular knowledge about how to live well with pigeons (Srinivasan 2019, 8). *How* to care for, support, tolerate, or even recognise the activities of nonhumans as claims, interjections, or experiments (van Dooren 2019; Wilson 2021) must still be negotiated each next time humans and nonhumans encounter one another in everyday life. What is necessary is greater attention to the ways humans and nonhumans do the work of 'processing' one another's activities as claims of various kinds (Barnett 2018, 8) and how people negotiate different modes of response and grammars of justification through their ongoing and everyday encounters. In 'leaving behind the notion of "norms" as controlling action and, instead, recovering them as local occasioned resources for "doing" order in context' (Ablitt and Smith 2019, 869), an EM study of encounters between humans and pigeons helps to recover how we learn with pigeons how to live with pigeons.

Conclusion

Rather than provoking constant problems or 'disorder' in everyday urban life, pigeons are familiar cohabitants of the city who are frequently ignored, overlooked, or unremarked upon—but who sometimes, to surprise, annoyance, or concern, show up 'out of place'. The disregard frequently shown towards pigeons should not be interpreted as the manifestation of an underlying 'cultural' or 'psychological' aversion or disdain, however. Such distance and disinterest can come to serve as a 'practice of accommodation' or 'vernacular knowledge about how to cohabit … as a response to non-human alterity': often enough, pigeons go 'visible but *not noticed*' (Srinivasan 2019, 10) as a way to share our streets with them, much as with other, human strangers.

In actual encounters between humans and pigeons, both distant curiosity and intimate confrontation can come to serve as techniques for living well in a multispecies city. By attending and responding to one another's activities, humans and pigeons alike contribute to a process of situated, embodied, and interactive public reasoning about how to get on together. Recognising and responding to the claims of nonhumans (Wilson 2021) is thus a situated challenge, involving conflict, disagreement, mistakes, and misinterpretation as well as curiosity, care, and kindness (Barnett 2018). Both an 'embodied empathy' for the experiences and lifeways of birds (Hodgetts and Lorimer 2020, 13) and a considered distance and recognition of limits of understanding (Giraud 2019; Rose et al. 2021) play a part in negotiating the terms of multispecies membership—how to respond, and *what it means to respond in that way*, is negotiated on the scene, not by culturally determined norms.

The ambivalence of encounters between pigeons and humans on trains thus supports neither a critical, nor an ameliorative style of analysis but rather one that seeks to open up a range of further questions about the terms and circumstances, the styles of reasoning, and repertoires of 'practices of accommodation' that are mobilised in order to learn how to share worlds with other animals. Perhaps, this points to another way of 'staying with the trouble'—to recall Donna Haraway's (2016, 10) frequently, if often too easily invoked aphorism—that does not focus solely on moments of conflict or extraordinary hybridity but attends, as well, to the everyday troubles humans and nonhumans go through to get along, past, around, or away from each other in day-to-day public life in a winged world—where flights, flocks, and feathers are *part* of the order of everyday life, if sometimes in unexpected ways.

Notes

1 The video 'Le quotidien d'un pigeon, METRO-BOULOT-IMPOT-DODO' was uploaded to youtube.com on 22 January 2017. Analytically, the video is, perhaps, *too* funny; it's useful for the argument developed in this chapter to pay attention to how and when the pigeon's 'daily grind' becomes seeable as funny to those in the video: https://www.youtube.com/watch?v=BASHSL5gsnk
2 For a trove of interesting examples, and an entertaining afternoon, see contributions to the sub-reddit r/birdstakingthetrain: https://www.reddit.com/r/birdstakingthetrain/
3 The video 'Pigeon entered in Delhi Metro at station and rescued by passengers at Next station.' was uploaded to youtube.com on 31 March 2018. While discussion here necessarily focuses on a short interaction, the video nicely shows a wide range of responses as those involved try to figure out what to do for themselves: https://www.youtube.com/shorts/Zw_wrUWfl6k

References

Ablitt, Jonathan and Robin James Smith. "Working out Douglas's aphorism: Discarded objects, categorisation practices, and moral inquiries." *The Sociological Review* 67, no. 4 (2019): 866–885. https://doi.org/10.1177/0038026119854271

Amir, Fahim. "Rats with wings." *Eurozine*. 30 August 2013. https://www.eurozine.com/rats-with-wings/

Arathoon, Jamie. "The geographies of care and training in the development of assistance dog partnerships." PhD thesis. University of Glasgow, 2022. https://doi.org/10.5525/gla.thesis.82798

Barnett, Clive. "Political affects in public space: Normative blind-spots in non-representational ontologies." *Transactions of the Institute of British Geographers* 33, no. 2 (2008): 186–200. https://doi.org/10.1111/j.1475-5661.2008.00298.x

Barnett, Clive. "Geography and ethics III: From moral geographies to geographies of worth." *Progress in Human Geography* 38, no. 1 (2014): 151–160. https://doi.org/10.1177%2F0309132513514708

Barnett, Clive. (2017). *The priority of injustice: Locating democracy in critical theory*. Athens, GA: University of Georgia Press, 2017.

Barnett, Clive. "Geography and the priority of injustice." *Annals of the American Association of Geographers* 108, no. 2 (2018): 317–326. https://doi.org/10.1080/24694452.2017.1365581

Barnett, Clive, Paul Cloke, Nick Clarke and Alice Malpass. *Globalizing responsibility: The political rationalities of ethical consumption*. Hoboken: John Wiley & Sons, 2011.

Bennett, Jane. *Vibrant matter: A political ecology of things*. Durham: Duke University Press, 2010.

Blechman, Andrew. "Flying rats." In *Trash animals: How we live with nature's filthy, feral, invasive, and unwanted species*, edited by Kelsi Nagy and Phillip David Johnson, 221–242. Minneapolis: University of Minnesota Press, 2013.

Bodden, Shawn. "A work-in-progress politics of space: Activist projects and the negotiation of throwntogetherness within the hostile environment of Hungarian politics." *City* 26, no. 2–3 (2022): 397–410. https://doi.org/10.1080/13604813.2022.2055930

Douglas, Mary. *Purity and danger: An analysis of concepts of pollution and taboo.* London: Routledge & Kegan Paul, 1984.

Escobar, Maria Paula. "The power of (dis)placement: Pigeons and urban regeneration in Trafalgar Square." *Cultural Geographies* 21, no. 3 (2014): 363–387. https://doi.org/10.1177/1474474013500223

Francis, David and Sally Hester. "Stephen Hester on the problem of culturalism." *Journal of Pragmatics* 118, no. 56–63 (2017). https://doi.org/10.1016/j.pragma.2017.05.005

Garfinkel, Harold. *Studies in ethnomethodology.* Hoboken: Prentice-Hall, 1967.

Garfinkel, Harold. *Ethnomethodology's program: Working out Durkheim's aphorism.* Lanham: Rowman & Littlefield, 2002.

Ginn, Franklin. "Sticky lives: Slugs, detachment and more-than-human ethics in the garden." *Transactions of the Institute of British Geographers* 39, no. 4 (2014): 532–544.

Giraud, Eva Haifa. *What comes after entanglement?.* Durham: Duke University Press, 2019.

Giraud, Eva Haifa. "After the 'age of wreckers and exterminators?'" *Cultural Politics* 17, no. 1 (2021): 37–47. https://doi.org/10.1215/17432197-8797501

Giraud, Eva Haifa and Gregory Hollin. "Care, laboratory beagles and affective utopia." *Theory, Culture & Society* 33, no. 4 (2016): 27–49. https://doi.org/10.1177/026327641561968

Goode, David. *Playing with my dog katie an ethnomethodological study of dog-human interaction.* West Lafayette: Purdue University Press, 2007.

Griffiths, Huw, Ingrid Poulter, and David Sibley. "Feral cats in the city." In *Animal spaces, beastly places*, edited by Chris Philo and Chris Wilbert, 56–70. London and New York: Routledge, 2000.

Haraway, Donna. *Simians, cyborgs, and women.* New York: Routledge, 1991.

Haraway, Donna. *Staying with the trouble.* Durham: Duke University Press, 2016.

Hodgetts, Timothy and Jamie Lorimer. "Animals' mobilities." *Progress in Human Geography* 44, no. 1 (2020): 4–26. https://doi.org/10.1177/0309132518817829

Hollin, Gregory, Isla Forsyth, Eva Giraud and Tracey Potts. "(Dis)entangling barad: Materialisms and ethics." *Social Studies of Science* 47, no. 6 (2017): 918–941. https://doi.org/10.1177/0306312717728344

Howlett, Jack, David List, Anne Boston, Lorna Read, Peter L.G. Bateman, Sabiha Foster and Jim Brock. "Passenger pigeons." *New Scientist.* 30 September 1995. https://www.newscientist.com/letter/mg14719978-600-passenger-pigeons/

Jayyusi, Lena. "Values and moral judgement: Communicative praxis as moral order." In *Ethnomethodology and the human sciences*, edited by Graham Button, 227–251. Cambridge: Cambridge University Press, 1991. https://doi.org/10.1017/CBO9780511611827.011

Jerolmack, Colin. "How pigeons became rats: The cultural-spatial logic of problem animals." *Social Problems* 55, no. 1 (2008): 72–94. https://doi.org/10.1525/sp.2008.55.1.72

Jerolmack, Colin. *The global pigeon.* Chicago: University of Chicago Press, 2013.

Latour, Bruno. *We have never been modern.* Cambridge: Harvard University Press, 1993.

Laurier, Eric. "The graphic transcript: Poaching comic book grammar for inscribing the visual, spatial and temporal aspects of action." *Geography Compass* 8, no. 4 (2014): 235–248. https://doi.org/10.1111/gec3.12123

Laurier, Eric. "YouTube fragments of a video-tropic atlas." *Area* 48, no. 4 (2016): 488–495. https://doi.org/10.1111/area.12157

Laurier, Eric, Ramia Maze and John Lundin. "Putting the dog back in the park: Animal and human mind-in-action." *Mind, Culture, and Activity* 13, no. 1 (2006): 2–24. https://doi.org/10.1207/s15327884mca1301_2

Laurier, Eric and Chris Philo. "Cold shoulders and napkins handed Gestures of responsibility." *Transactions of the Institute of British Geographers* 31 (2006): 193–207. https://doi.org/10.1111/j.1475-5661.2006.00205.x

Law, John. *After method: Mess in social science research.* London and New York: Routledge, 2004.

Law, John and Vicky Singleton. "ANT and politics: Working in and on the world." *Qualitative Sociology* 36, no. 4 (2013): 485–502. https://doi.org/10.1007/s11133-013-9263-7

Leys, Ruth. *The ascent of affect: Genealogy and Critique.* Chicago: University of Chicago Press, 2017.

Lynch, Heather. "Esposito's affirmative biopolitics in multispecies homes." *European Journal of Social Theory* 22, no. 3 (2019): 364–81.

Lynch, Michael. "Revisiting the cultural dope." *Human Studies* 35, no. 2 (2012): 223–233. https://doi.org/10.1007/s10746-012-9227-z

Massey, Doreen. *For space.* London: Sage, 2005.

Mondada, Lorenza. "Multiple Temporalities of Language and Body in Interaction: Challenges for Transcribing Multimodality." *Research on Language and Social Interaction* 51, no. 1 (2018): 85–106. https://doi.org/10.1080/08351813.2018.1413878

Mosco, Rosemary. *A pocket guide to pigeon watching: Getting to know the world's most misunderstood bird.* New York: Workman Publishing, 2021.

Muñoz, Daniel. "Carrying the rollator together: A passenger with reduced mobility being assisted in public transport." *Gesprächsforschung - Online-Zeitschrift zur verbalen Interaktio.* 22 (2021a): 591–614.

Muñoz, Daniel. "Accessibility as a 'doing': The everyday production of Santiago de Chile's public transport system as an accessible infrastructure." *Landscape Research* (2021b): 1–12. https://doi.org/10.1080/01426397.2021.1961701

Philo, Chris. "Animals, geography, and the city: Notes on inclusions and exclusions." *Environment and Planning D: Society and Space* 13 (1995): 655–681.

Philo, Chris. "Less-than-human geographies." *Political Geography* 60 (2017): 256–258.

Philo, Chris. "Nothing-much geographies, or towards micrological investigations." *Geographische Zeitschrift* 109, no. 2–3 (2021): 73–95. https://doi.org/10.25162/gz-2021-0006

Philo, Chris and Chris Wilbert, eds. *Animal spaces, beastly places.* New York: Routledge, 2000.

Rose, Mitch, David Bisell and Paul Harrison. "Negative geographies." In *Negative geographies: Exploring the politics of limits,* edited by David Bisell, Mitch Rose and Paul Harrison, 1–38. Lincoln: University of Nebraska Press, 2021. https://doi.org/10.2307/j.ctv1z3hkh4

Savransky, Martin. "The pluralistic problematic: William James and the pragmatics of the pluriverse." *Theory, Culture & Society* 38, no. 2 (2021): 141–159. https://doi.org/10.1177/0263276419848030

Simms, Eric. *The public life of street pigeon.* London: Hutchinson, 1979.

Srinivasan, Krithika. "Remaking more-than-human society: Thought experiments on street dogs as 'nature'." *Transactions of the Institute of British Geographers* 44, no. 2 (2019): 376–391. https://doi.org/10.1111/tran.12291

Van Dooren, Thom. *Flight ways: Life and loss at the edge of extinction.* New York: Columbia University Press, 2014.

Van Dooren, Thom. *The wake of crows.* New York: Columbia University Press, 2019.

Whatmore, Sarah. *Hybrid geographies: Natures cultures spaces.* London: Sage, 2002.

Wilson, Helen F. "Passing propinquities in the multicultural city: The everyday encounters of bus passengering." *Environment and Planning A: Economy and Space* 43, no. 3 (2011): 634–649. https://doi.org/10.1068/a43354

Wilson, Helen F. "On geography and encounter: Bodies, borders, and difference." *Progress in Human Geography* 41, no. 4 (2017): 451–471. https://doi.org/10.1177/0309132516645958

Wilson, Helen F. "Seabirds in the city: Urban futures and fraught coexistence." *Transactions of the Institute of British Geographers*. Early View. (2021). https://doi.org/10.1111/tran.12525

PART III

Flights of fancy

9

BIRDS AS WINGED WORDS

A reading of Aristophanes, *The Birds*

Jeremy Mynott

Aristophanes is now the best known of the ancient Greek comic dramatists. He lived in Athens in the fifth century BC and wrote over forty comedies, of which just eleven survive. His plays are all written to be performed as broad farce, usually laced with plenty of obscenity, blasphemy, and slanderous parodies of local celebrities and public figures; but at a deeper level, the ideas in them are radical and subversive, both socially and politically.[1] In the case of *The Birds*, one might also add 'taxonomically' subversive, since a key theme in the play is the way birds, humans, and gods keep exchanging roles in the power structure.

The basic plot of *The Birds* is an escapist fantasy. Very briefly, it goes like this. Two Athenian citizens, Peisetaerus ('persuader of his comrade') and Euelpides ('son of good hope'), are fed up with high taxes, over-regulation, and life in the city generally. They negotiate with the leader of the birds, a hoopoe, and his chorus of twenty-four other species of birds to establish a new city in the sky, which they propose to call *Nephelo-kokkugia* (that is, 'Cloud-Cuckoo Land'). From this intermediate kingdom, poised between earth and heaven, they plan to establish border controls and start charging the gods a tariff on the smoke of sacrifices passing up from earth to the upper kingdom of the gods. They persuade the birds to build a Great Wall (a move with some modern political resonances) to keep the gods out and enforce the policy. The venture is a great success and all their fellow Athenians go 'ornitho-manic' with excitement (that's Aristophanes' word) and rush out to buy wings to join the two pioneers. The play ends with the surrender of the gods, whose power passes to the birds and their human benefactors.

Now, at one level, that would have been played, and enjoyed by the audience, as pure entertainment – with plenty of coarse jokes, satirical references to people

DOI: 10.4324/9781003334767-13

in the news, and featuring actors in colourful bird costumes pretending to be birds, who were in turn behaving and talking like humans.

At another level, *The Birds* would probably also have been seen as a larger, and darker, political satire. The play was produced in 414 BC, just at the time of the vastly over-ambitious Athenian expedition against Sicily, which turned out to be an act of imperial *hubris* that ended in military disaster and effectively lost Athens the protracted war they had been fighting against Sparta since 432 BC. The timing of Aristophanes' play imagining another kind of imperial fantasy – that is, colonising the skies – can scarcely be accidental. This too perhaps has some contemporary resonance, in the aspiration to found colonies on the moon to escape the threats to the earth posed by the climate crisis we have ourselves inflicted on our environment.

But the third level at which we can respond to the play is as an exploration of the idea that birds have a special significance as messengers or intermediaries and that's what I focus on here:

Right at the start, the Athenian adventurers choose two corvids, a jackdaw and a crow, as their guides to help them infiltrate the kingdom of the birds (1ff.). Corvids were probably selected by Aristophanes for this role because of their familiarity and for their well-known intelligence and mimetic powers.

The jackdaw and crow introduce the two Athenians to the leader of the birds, a hoopoe, who is able to act as interpreter between the human interlopers and the birds they want to negotiate with because, he says, he was once a human and 'understands everything a human does and everything a bird does too' (119–20). He's also taught the other birds to speak some Greek, 'I have lived with the birds a long time and they are no longer the barbarians they were before I taught them to speak properly' (199–200).

We then have some code-switching. The hoopoe first mimics himself in birdspeak with an extended *epopopopopopopoi*; he then does an impression of the nightingale he has just coaxed out of the thicket *io, io, ito, ito, ito, ito, ito*; and, in what must have been a virtuoso performance on stage, he next summons the rest of the birds out of hiding from their different named habitats in a whole series of imitations *tio tio tio tio tio tio ti tio; trioto trioto totbrix; attagen attagen; torotorotorotorotix; kikkabu*; and *torotortorolililix* (227–63). In Aristophanes' transcriptions of these, I can reliably identify only the nightingale, little owl, and one or two other onomatopoeic songsters, but we should remember that the songs and calls will have been composed at least partly to fit in with the demands of the metre and the accompanying music. That music would have been more important and integral to the action than music generally is in modern theatre. Those present were literally an *audience*, not just spectators. Remember too that these dramatic performances took place in open-air theatres. *The Birds* was first produced at the festival known as the 'City Dionysia' in spring (late March to early April) 414 BC, at the Theatre of Dionysus just south of the Acropolis. There would have been bird song everywhere, clearly audible.

The twenty-four birds in the chorus supporting the hoopoe (*epops*) are all named in turn (297–304). Most are clearly identifiable (and would have been recognised by the Athenian audience) and are likely to have been selected at least partly because of the opportunities offered to the costume-makers by their varied and colourful plumages:[2]

Perdix partridge
Attagas francolin
Penelops wigeon
Alcyon kingfisher
Keirulos ? sea bird
Glaux owl
Kitta jay
Trugon turtle dove
Korydos lark
Eleas warbler
Hypothumis ? wheatear
Peristera rock dove
Nertos vulture
Hierax hawk
Phatta wood pigeon
Kokkyx cuckoo
Erythropous ? redshank
Keblypuris ? woodchat shrike
Porphuris purple gallinule
Kerkines kestrel
Kolymbis little grebe
Ampelis ? bunting
Phene lammergeier
Druops woodpecker

Later in the play, the chorus of birds stress their value to humans, first as a marker of seasonal change:

All the greatest blessings mankind enjoys comes from us, the birds.
For a start, we reveal the seasons of spring, winter and autumn.
The time to sow is when the crane flies off to Libya, bugling:
That tells the sailor to hang up his rudder and take a lie-in …
Next it's the kite's turn to arrive and herald another season;
This is the time for your spring sheep-shearing.
Then it's the swallow – time to sell those winter woollens
And get yourself something more summery.

(708–15)

Secondly, they describe themselves as special consultants. On reading this passage in translation, parts of it won't seem to make much sense initially:

> We are your oracles ...
> You don't start on anything without first consulting the birds,
> Whether it's about business affairs, making a living, or getting married.
> Every prophecy that involves a decision, you classify as a bird.
> To you, a significant remark is a bird; you call a sneeze a bird,
> A chance meeting is a bird, a sudden sound, a servant or a donkey – all birds.
> So clearly, we are your gods of prophecy.
>
> (716–24)

That sounds like some bizarre mistranslation, until you remember that the Greek word for a bird, *ornis*, was also their word for an omen. That's not a coincidental fact. These are not just two homonyms, words which look and sound the same but have quite different meanings, like the bark of a tree and the bark of a dog. They are what linguists call polysemes, which are connected in a deeper, metaphorical way, like the mouth of a river and the human mouth. Birds were signs and acted as the agents through which the gods could convey messages to humankind. So, all those things called 'birds' in the extract I just read, turn out to involve common superstitions, rather like our habit of saying 'bless you' when someone sneezes. To call something a 'bird' was to say it might be significant and that birds might be the clue to what that significance was.

Birds, that is, to adapt the famous remark of Lévi-Strauss (1962), were 'good to think with'. And this connects to the larger plot of the play. Aristophanes said that the citizens of Athens went *ornitho-manic* in their excitement to join the birds in the sky, but they also went *ornitho-morphic* in dressing up like birds and taking on their characteristics:

> They're all bird-mad now and are only too pleased
> To imitate everything the birds do.
> They rise with the lark from their beds
> And fly down to enjoy a diet of laws;
> Then they brood on their books
> And peck away at the bills.
> And the height of this ornitho-mania is
> That many of them get given bird-names too.
> [*various untranslatable puns of unknown local characters follow*]
> And these bird-lovers are all singing songs
> That mention birds – a swallow or wigeon or goose
> Or a pigeon – or wings, or even a shred of feather.
>
> (1283–302)

FIGURE 9.1 Bird-dancers.

Meanwhile the birds themselves were being *anthropo-morphised* when they were described as turning the tables on the humans who had traditionally hunted, eaten, caged, and abused them:

> They treat you like slaves, foreigners and fools
> ... even in temples, every bird-hunter is after you,
> Setting up nooses, snares, sticky twigs,
> Traps, nets, toils and decoys.
> And when they've caught you they sell you whole;
> The customers feel up your flesh,
> And if they go ahead they don't just
> Roast you and serve you up,
> But they coat you with cheese ...
> Prepare a sweet, greasy dressing
> And pour it all over you hot.
>
> (523–35)

FIGURE 9.2 The Cambridge Greek Play, 1903: Aristophanes, *The Birds*. The splendidly costumed cast of *The Birds*, performed 'in the original Greek' at the New Theatre in Cambridge in 1903. The chorus of twenty-four different birds is arranged either side of the two Athenian visitors to Cloud-Cuckoo-Land with Tereus, the hoopoe, behind them. The 'Cambridge Greek Play' is a famous tradition, begun in 1882, which continues to the present day. The early performances enjoyed great prestige as part of the nineteenth-century enthusiasm for Hellenic culture, and attendance at them clearly had some social cachet – a special train was put on from King's Cross and back for performances of the inaugural 1882 play (Sophocles, *Ajax*). The 1903 performance of *The Birds* was accompanied by music from Sir Hubert Parry and was attended by figures like Prince Albert Victor and the Postmaster General. It attracted very favourable reviews in the national press (though *The Times* sniffily complained that some members of the audience who didn't understand Greek were laughing in the wrong places); in the provincial press (including the *Yorkshire Daily Post*, the *Liverpool Post*, and the *Blackburn Weekly Standard*, which gave it two and a half columns); and even internationally (in *Le Figaro*). *The Birds* remained a regular favourite and was performed in 1883 (probably the first complete production since antiquity), 1903 (as illustrated here), 1924, 1971, and 1995, its success evidence both of the colourful dramatic possibilities the play offers and the continuing fascination with its avian fantasies and embedded symbolisms.

Source: The Syndics of Cambridge University Library.

In revenge, the birds warn any humans who keeps caged birds to release them, or else:

> If you disobey, you'll be arrested by the birds
> And it will be your turn to be tied up and serve as decoys.
>
> (1085–87)

Or they threaten to abuse them in other humiliating ways:

> And if you don't support us …
> Then the next time you're wearing your best suit
> We'll make you pay the penalty –
> With all the birds crapping on you!
>
> (1114–17)

After the birds complete the construction of Cloud-Cuckoo Land, however, the humans can't wait to join them, even to the point of adopting prosthetic aids:

> That's the news from Earth, and I can tell you this:
> There will soon be thousands of them heading this way,
> All wanting wings and the kit for living like birds.
> So, you'd better lay in supplies for a wave of immigrants.
>
> (1304–7)

★

The play as a whole is a *locus classicus* for the symbolic potential of birds, the first extended treatment of its kind in Western European literature. It drew to some extent on an earlier folkloric tradition of animal fables, like those of Aesop, where birds and other animals are given representative human traits and enact moralising stories like 'The eagle and the wren', 'The fox and the grapes', 'The crow and the water-jar', and so on. *The Birds* has many resonances, too, in later works that seek to model the realm of birds on human society. Chaucer's *Parlement of Foules* (c. 1383), for example, also features birds behaving and speaking just like humans, and as in the Aristophanes play, he also has them code-switch between the two languages, so playing on the ancient debate about whether the capacity for speech was what defined rationality, the usual criterion of human exceptionalism.[3]

Indeed, the impulse to anthropomorphise animal society remained so strong that it extended even to popular understandings of subsequent scientific taxonomies. In his classic work, *Man and the Natural* World, Keith Thomas describes how the Linnaean system, as propounded in a late eighteenth-century English edition, pressed the parallels between the natural world and human society very closely indeed:

> The 'Vegetable Kingdom' was divided into 'Tribes' and 'Nations', the latter bearing titles which were more sociological than botanical: the grasses were 'plebeians' – 'the more they are taxed and trod upon, the more they multiply'; the lilies were 'patricians' – 'they amuse the eye and adorn the vegetable kingdom with the splendor of courts'; the mosses were 'servants' – who 'collect for the benefit of others the daedal soil'; the flags were 'slaves' – 'squalid, revivescent, abstemious, almost naked'; and the funguses were 'vagabonds' – 'barbarous, naked, putrescent, rapacious, voracious'.[4]

And the hierarchical assumption that some forms of life are 'lower' than others persists as strongly today as it did in Aristotle's pioneering zoological taxonomy in the fourth century BC.

★

Why then, did Aristophanes particularly choose birds as the principal characters in this play and the vehicles through which to perform this exploration of our relationships with the natural world? After all, *The Birds* is not the only ancient Greek play whose title references animal species. Aristophanes himself wrote plays entitled *The Frogs* and *The Wasps*, while lost plays by other authors include the *Ant-men, Bees, Goats, Fishes,* and even the *Fruit Flies*.

The sheer physical ubiquity and abundance of birds must have been one factor. There they were – visible and audible at most times of the day; occupying all the domains of land, air, and sea; and common in urban, cultivated, and wild areas alike. It is hard for us now to imagine the natural abundance of the bird life that is featured so prominently in the literature and art of the ancient world. Kites, corvids, pigeons, and sparrows were regular city scavengers; owls, swallows, martins, and swifts nested in temples and public buildings; nightingales could be heard singing in the suburbs of Athens; there were cuckoos, hoopoes, warblers, and woodpeckers within city limits; while eagles and vultures would have been a common sight overhead in the countryside beyond. People will have heard or seen a bird every day of their lives. Mammals by contrast were both less numerous and less conspicuous, more likely to be shy or nocturnal, and always earth-bound. Fish were found in just one domain, while the multitudes of insects were not much differentiated and were less easy to relate to in the same way.[5]

Birds are also on a convenient physical scale – large enough to be easy to see clearly and small enough to act as convenient miniature models for human comparisons. And in structural terms, the similarities are striking. Like humans, birds stand upright on two legs and have rounded heads with prominent eyes. No surprise then that owls and penguins feature so commonly in the soft toys department. Moreover, birds can walk, run, hop, jump, wade, and paddle, just as we can.

Birds were an active presence in the world, too. They were prominent in the foreground of people's experience, interacting with them as both parties went about their daily lives. They used mankind's buildings for nest sites and perches; they scavenged around their towns and villages; they fed in their fields and predated their crops; they followed armies on the move and fishermen in their ships; and in some cases, they even shared human dwellings as family pets and companions.

Bird behaviour also provides ample examples and analogues for a sympathetic rapport, now as then. We observe their social gatherings and can watch them eating, playing, courting, displaying, homemaking, competing, and fighting. We can hear them singing, communicating, and communing. We think we know what they are up to and can relate to their perceived purposes and activities.

Bird vocalisations in particular have always been important indices of our kinship with birds. We can sense their meanings, at least to the extent of distinguishing between alarm calls, flocking and 'conversational' calls, the begging calls of chicks, and the mating and seasonal songs of adults, for example. Our responses to and celebration of bird song permeate every world literature, beginning with Greek and Latin, and may underlie much human music too, as the Roman poet Lucretius speculates:

Men learned to imitate by mouth the liquid notes
Of birds long before they could join together
Singing tuneful songs to delight the ear.
(Lucretius, *De Rerum Natura* V, 1379–81)

Above all, birds have the gift of flight, a perennial topic of human fascination, and at some level, a deep human aspiration too, which has been expressed in all kinds of metaphorical and symbolic ways, from the Icarus myth through to Sigmund Freud's interpretations of flying dreams. 'Winged birds' was the standard Homeric phrase, which captures this identification of birds with the power of flight. Aristophanes frequently uses the same epithet and the chorus of birds in the play reminds humans of their lowly earth-bound condition:

Enfeebled by nature, just like the kinds of leaves,
Weaklings, formed from clay, a shadowy race of puny beings,
Flightless creatures of a day, wretched mortals, fleeting as dreams.
(685–87)

Birds by contrast were creatures of the air, the realm intermediate between humans and gods in which they moved with such ease. Angels are awarded wings for good reason.[6]

Wing-less birds like ostriches were regarded as anomalies – 'dualisers' Aristotle called them, which crossed zoological categories and seemed to have more in common with big game. Mammals, insects, flowers, trees, and fish were all important presences in the landscape, too, but they each lacked in some way this combination of characteristics that made birds seem such suitable subjects for stories, myths, fables, allegories, and other human imaginings. That in the end was surely why birds featured so strongly as omens and metaphorical messengers in the ancient world. They were ideally constructed, both physically and symbolically, to be the bearers of signs and meanings.

★

In the climax to *The Birds*, Aristophanes describes how everyone is now asking for a set of wings in order to join the birds in their new city in the skies. The comic hero, Peisetaerus, is the man in charge of the supplies, and he admonishes the last applicant, an Informer, who wants wings just to fly round the cities more quickly on his dubious legal business. Aristophanes then has a riff on the idea that language itself has the power to 'take wing':

INFORMER: I don't want your advice, good sir, I just want wings.
PEISETAERUS: That's just what I'm trying to give you now, through my words.
I: How can you give a man wings through words?
P: Didn't you know? Words can help everyone take wing ...

Peisetaerus gives him some examples of winged hopes and ambitions, and points out that drama itself can 'cause the heart to flutter'. He then concludes:

Yes, it's by words that the mind is uplifted
and mankind soars aloft.

(1436–49)

Flight was the prerogative of birds and a good part of their fascination for earthbound mortals. Language, meanwhile, was supposed to be the unique distinguishing feature of human kind. But here, in an artful metaphor, Aristophanes forges the connection between the two faculties and with them the two realms of birds and humanity, each of which has been penetrated by the other.

Notes

1 They range, for example, over: democracy and demagogy (*Knights*), militarism (*Acharnians* and *Peace*), the legal system (*Wasps*), education and sophistry (*Clouds*), culture and tradition (*Frogs*), and sexual politics (*Lysistrata* and *Women in Parliament*).
2 On the costumes, see Dunbar (1995), 244; on the names and identifications, see Mynott (2018), 263–65. I have added question marks in the case of more speculative identifications.
3 See Warren (2018), 147–68 and Sax (2021), 161–63.
4 Thomas (1983), 66.
5 With the honorific exception of bees and ants, which were regularly admired for their social organisation.
6 See further: Mynott (2009), 283–85; Mynott (2018), 289–94; and Pope (2021), 147–63.

References

Dunbar, Nan, ed. 1995. *Aristophanes: birds* (Oxford University Press, Oxford).
Lévi-Strauss, Claude. 1962. *Le Totémisme aujourdhui* (Presses Universitaires de France, Paris).
Mynott, Jeremy. 2009. *Birdscapes: birds in our imagination and experience* (Princeton University Press, Princeton, New Jersey).
Mynott, Jeremy. 2018. *Birds in the Ancient World: winged words* (Oxford University Press, Oxford).
Pope, Richard. 2021. *Flight from grace: a cultural history of humans and birds* (McGill-Queen's University Press, Monteal & Kingston).

Sax, Boria. 2021. *Avian Illuminations: a cultural history of birds* (Reaktion Books, London).

Thomas, Keith. 1983. *Man and the Natural World: changing attitudes in England, 1500–1800* (Allen Lane, London).

Warren, Michael. 2018. *Birds in Medieval English Poetry: metaphors, realities, transformations* (D.S. Brewer, Cambridge).

10

BIRDS AND CHRISTIAN IMAGERY

Roger S. Wotton

Introduction

I taught Animal Form and Function at UCL (University College London), with lectures based on a traditional approach that would have been familiar to nineteenth-century natural historians but brought up to date with the latest information. Practical classes followed a similar tradition in that we looked at both living material and preserved specimens, and skeletons, from UCL's wonderful Grant Museum of Zoology. We also conducted experiments and made dissections, and I was very pleased that students told me how much they enjoyed it all.

Animal locomotion was an important part of the course, and we started by looking at movement in water, followed by the invasion of land, and then on to the conquest of the air. That followed an evolutionary sequence, with flight evolving independently in insects, birds, and bats, and in looking at birds, one is conscious of two features: the covering of feathers over the body and the presence of wings. Bird wings are shaped by feathers that are very light and yet strong; other feathers producing a streamlined profile, or being used for insulation, as birds have a high body temperature (higher than that of mammals), unlike almost all their reptilian ancestors. Streamlining reduces pressure drag by encouraging a delay in separation point of airflow over the body, and this produces a smaller wake. All this is fascinating, but it doesn't lead us into the subject of this chapter.

As well as my interest in biology, I have always been fascinated by paintings and sculpture and the insight that they give us into cultural history. My approach to the study of the visual arts is influenced by my training as a biologist and a scientific background also brings with it a certain scepticism. An example comes when artists create impressions of dinosaurs, something with which we are all familiar. Descriptions of fossil reptile bones are often accompanied by a painting of the whole animal in a landscape that "brings it to life" and these images we

DOI: 10.4324/9781003334767-14

believe to be real, especially when they are reinforced by repetition. Christian imagery can be like that, too, as we shall see.

Wanting to fly

We envy birds' ability to fly and, although aeroplanes enable us to cover large distances at high speed, we need an external means of power, and artificial wings, to do so. Perhaps, we come closest to bird flight in gliders that have an airframe, control surfaces, and long wings, and mimic birds in gliding and soaring flight, with pilots using thermals and air currents to remain airborne for long periods. However, gliders must be launched from the ground to a safe altitude using a cable, or a tug aircraft, before making their controlled descent. Flapping flight is impossible.

Another form of human flight is enjoyed by wingsuit flyers who jump from cliffs, or aircraft, and then use movements of their arms and legs to control a fabric suit that is extended to resemble the skin flaps (patagia) of flying squirrels. This parallel is strong, as the squirrels run up tree trunks before launching themselves through the air and then landing on another tree some distance away, producing changes in direction by movements of the limbs that consequently change the movement of air over the skin flaps. Using the same principles, wingsuit flyers can fly very fast, and close to terrain, but they must eventually open a parachute to ensure a soft landing. It is not much like bird flight.

Birds and imagery

Since our earliest appearance, humans have observed birds and learned that they vary widely in size and in feeding habits. We also watched as they flew rapidly over large distances and ascended high into the sky, often to the limits of human vision.

Our envy of birds, and their ability to fly, has given them an important role in mythology and imagery, and birds, and bird wings, have been used to suggest flight to other worlds. This ability not only applies to birds, as some fairies and dragons are also winged and inhabit imaginary places from which they come to visit us. Of those that fly, fairies have either damselfly, or butterfly, wings, and dragons those of bats, not those of birds (Wotton 2009).

The use of birds in Christian imagery stems from The Bible, with the New Testament being added to the Old Testament of the Jewish faith. Modern Christian images have a lineage that is many centuries old, and mediaeval and Renaissance paintings, carvings, and statues have an enduring influence. At the time they were produced, we still had a geocentric view of the universe, with the planets, and the Sun, revolving around the Earth in concentric orbits, with stars in a ring beyond the planets. We had little idea how far away the planets and stars were and could easily believe that Heaven must be somewhere just beyond the stars, with Hell being somewhere within the Earth, with fiery volcanic eruptions indicating its location.

With this background in mind, we can turn to images of birds that have been adopted by the Christian tradition and by secular culture. Many birds are used as symbols, but I confine myself to certain key taxa that have the greatest symbolic importance (Ferguson 1961).

Eagles

Eagles have always been admired for their hunting ability, their large size, and (to our eyes) their fierce appearance. These attributes have led to their use as symbols of power. In addition to their representation on contemporary national flags, buildings, and monuments, eagles (*Aquila*) were important in Roman times, where aquilifers carried the eagle staffs that symbolised Roman legions. Given the time of the writing of the New Testament, it therefore comes as no surprise to find that the eagle was a symbol of power, and not just in a military sense. Visitors to churches and cathedrals of the Catholic tradition are familiar with eagle lecterns used for reading from The Bible that was held on the eagle's back. Some are very large and imposing.[1]

Eagles had other symbolism as Ferguson (1961) writes:

> The eagle may generally be interpreted as a symbol of the Resurrection. This is based upon the early belief that the eagle, unlike other birds, periodically renewed its plumage and its youth by flying near the sun and then plunging into the water.

Ferguson continues by describing the ability of eagles to soar so high that they cannot be seen by the naked eye and that they were thought to have the capacity to look at the Sun. These attributes have led eagles to be used as symbols of Christ and of the power of the Gospels, especially the Gospel of St John, with whom the eagle is often associated (Ferguson 1961).

Eagles are not the only birds of prey used in Christian imagery as vultures are shown in some paintings and they are well known for feeding on carrion, making them symbols of impending death. The same is true of large black birds like crows and ravens.

Vultures and large black birds

In *The Agony in the Garden* (Figure 10.1) painted by Andrea Mantegna in 1455–6, we see Christ praying while, in the foreground, his disciples sleep. In the distance, Judas is leading a group of people, including soldiers who are coming to arrest Jesus, and an air of foreboding is provided by the vulture sitting in the tree. In addition to the vulture, we see two groups of rabbits, both moving away, and, while they are outside the topic of this chapter, we know that these represent people who have put their faith in Christ (Ferguson 1961). It makes the static vulture an even more potent symbol of what is to come, and other painters have used the same device in religious paintings.

The image of a large black bird, or a flock of black birds, is also found in paintings on secular themes as an allusion to the threat of death. Examples are the crows

FIGURE 10.1 Andrea Mantegna "The Agony in the Garden."

seen in works by Akseli Gallen-Kallela and Vincent van Gogh.[2] In the former, a very poor boy in patched clothes looks across part of a meadow on which his family is heavily dependent, while a crow pecks at the ground and looks at him. A simple, yet powerful, image that tells a story of hardship and survival. Van Gogh's *Wheatfield with Crows* gives the viewer a choice of roads to walk down; the central one leading to an invisible ending. Flying towards us in a dark sky are many crows, and we are left with the bleak view that this road leads to the prospect of death. This interpretation becomes even more powerful when one realises that van Gogh committed suicide shortly after completing this work.

In the northern latitudes where both Gallen-Kallela and van Gogh lived, crows are common and vultures rare, so the latter are very unlikely to appear in their paintings. Both types of birds convey the same message and it is quite different to the symbolism afforded by doves. These are perhaps the commonest birds in Christian imagery, and they are used as symbols in various ways.

Doves

Doves have been domesticated for thousands of years, after wild rock doves (*Columba livia*) that nest on rocky cliffs were captured and bred for food and then became feral, producing the familiar pigeons of parks and squares in cities. Other selective breeding of rock doves produced homing pigeons, with extraordinary powers of return to their "home" should they be displaced, a habit that has resulted in sporting contests and their use for conveying messages back to base from distant locations. The descendants of these domesticated rock doves come in many colour morphs, including the pigeons that we call white doves that are bred in large numbers and which are released as symbols of peace. Unfortunately, these releases are a gift for birds of prey, and some "peaceful" white doves may suffer a grisly fate a short time after being released.

The first use of doves in Christian imagery comes from the book of Genesis in The Bible (and my source is the Authorised [King James] Bible and is, of course, dependent on the exegesis of biblical scholars). In Chapter 8, we read that Noah, having sailed in the ark to avoid the great flood, released a dove "to see if the waters were abated from off the face of the ground." The dove came back as expected, and Noah repeated the process after a further seven days. This time the dove returned with an olive twig in its beak, and Noah thus knew that the flood was receding and that his family, and all the animals in the ark, would now be safe. White doves bearing an olive twig are an enduring symbol of peace, although not everyone is aware of the origins of the image.

White doves are also used to represent the Holy Spirit, a "being" that is impossible to visualise, except in symbolic terms. There is reference to this in St Matthew's gospel: ".. Jesus, when he was baptised, went up straightaway out of the water: and lo, the heavens were opened unto him, and he saw the Spirit of God descending like a dove, and lighting upon him." Piero's conception of the baptism[3] shows the white dove descending on widely spread wings to hover above Christ's head, and the exquisite painting by Murillo of *The Heavenly and Earthly Trinities* (Figure 10.2) shows God in Heaven and the young Jesus on Earth, with a dove providing the third link in the Heavenly Trinity. The Earthly Trinity of Jesus, Mary, and Joseph is handled with great tenderness; His parents holding, and supporting, His hands while He stands above them on what could be a mounting block. The imagery is very powerful and the dove, surrounded by a corona, is in the middle of it.

If the Holy Spirit presents a challenge to painters and other creators of imagery, so do angels. Like the Holy Spirit, angels connect Heaven and Earth, and we can all identify them – even young children know what an angel looks like and they feature in their nativity plays each Christmas, as well as in countless stories.

Angels

Angels, whether in Heaven or on Earth are androgynous, robed human figures that, almost universally, have arms and also bird wings that are attached to their backs. I say almost universally because there are several paintings of angels, like those by Piero and Rosetti where they do not have wings.[4] If we look at the painting by Piero closely, we can identify key characters in the story of the Nativity, but the angels confuse us – they are playing and singing and glorifying the new-born Christ, but we're not entirely sure that they are angels, as their wings are missing or are not in view.

FIGURE 10.2 Bartolomé Esteban Murillo "The Heavenly and Earthly Trinities."

The familiar image of winged angels emanated from fourth-century Christianity, probably using classical statues of winged goddesses as models. That they were bird wings comes from our observations of birds' power of flight and our envy of this ability. But why this accepted image? In a fascinating essay, Therese Martin (2001) explains that the earliest angels in Christian imagery were not winged, were also referred to as men (this was used as an interchangeable term), and were wrapped in brilliant light. Some wore beards and "at least one is balding," and Martin explains the transition from male to androgynous, winged angels was because "angels as men were too attractive to women – and too susceptible to their female charms" (Martin 2001). She concludes that:

> ..the image of the winged angel manages to combine a female model with a male nominative ("angelos") in order to portray a creature that is spirit, not belonging to either gender. The visual form of a man with wings satisfied the perceived need in the fourth century for an angel that was neither man nor woman, neither man nor God. The winged image was a successful visual metaphor that managed to capture all the ambiguities inherent in the angelic nature and explain them in a way that has remained clearly convincing to the present day.
>
> *(Martin 2001)*

FIGURE 10.3   Francesco Botticini "The Assumption of the Virgin."

Many paintings show angels and we note that, while the general appearance of angels has not changed, there have been differences in detail, one of which comes in the colour of their wings. In Botticini's late fifteenth-century painting of *The Assumption of the Virgin* (Figure 10.3), we see angels in a Heaven that appears as a dome above the rings of the geocentric universe – it looks very close compared to the distant horizon.

Let's examine this powerful image in more detail. In Heaven, we recognise Christ and Mary, surrounded by rows of angels all having colour-coded robes and wings, each row representing a category of angel as recognised by Pseudo-Dionysius the Areopagite and accepted by the Catholic Church. In ascending order, these are:

Angels – pink robes: gold and black wings
Archangels – white robes: pink wings (all holding urns) (It is accepted that there are three archangels, but there is reference to others in religious texts – the number shown here is unusual).
Principalities – blue robes: black and red wings

Dominions – olive robes: red wings
Virtues – blue robes: blue and red wings
Powers – white robes: pink and olive wings

Thrones, Cherubim, and Seraphim are shown as babies (cherubs) – although The Bible describes Seraphim as having six wings and Cherubim having four wings. Is there any evidence for as many Seraphim and Cherubim as Botticini has shown?

All these angelic beings have their roles to play, but only Angels and Archangels leave Heaven to visit Earth, so "need" their wings. At first glance, we miss the human figures shown in the painting and perhaps confuse them with the angelic throng. Seated between the ranks of angels are women and both bearded, and clean-shaven, men. Many are clothed in monastic, or ecclesiastical, garb and have gained entry to Heaven as Saints, Old Testament figures, and persons of similar status.

One of the features of Botticini's angels that strikes us is their vivid colour, and we often forget that places of Christian worship in mediaeval times were also brightly coloured, a contrast to the interiors of most churches and cathedrals today. Many other paintings from this period show angels with brightly coloured wings, but the Renaissance saw a change to portraying angels with white, or very pale, bird wings, often shown in accurate anatomical detail,[5] a tradition that has continued.[6]

Representations of angels in modern art maintain the same basic structure, but there is an exception – Antony Gormley's *The Angel of the North*.[7] This huge work has wings outstretched, and there are no separate arms, as shown in most images of angels. There are also no obvious flowing robes, although we sense their presence.

As angels are shown as physical beings in images, we are left with the question of whether angels generally exist in this form. We know that Jesus had a human body and that He ascended into Heaven. Perhaps, angels only appear as physical beings when they need to, but why then are they shown as having bird wings? It is quite different to the representation of the Holy Spirit as a dove, as we know that this part of the Trinity cannot be represented in any other way. There are many that believe not only that angels exist but also that they exist in the form in which they are shown in paintings and sculptures. There is belief that they can fly between Heaven and Earth and that their wings are used in some way during this flight. Before we consider this further, we need to know more about the structure of birds and how they fly.

Birds and flight

It is accepted that birds evolved from reptilian ancestors and that their wings developed from forelimbs. The loss of dexterity was then compensated by the development of the beak: different shapes of beak evolving for different purposes. With very few exceptions, birds fly and having a light skeleton is an important prerequisite. The skeleton of a pheasant is shown in link 8[8], and these ground-dwelling birds use flight as a means of escape from predators (and beaters in shooting parties!). When disturbed, pheasants power upwards, with much noise from the wings clattering together.

Several features of the bird skeleton should be emphasised. Firstly, there are many modifications that ensure its lightness, with a reduction in the size of some bones, and bracing between adjacent ribs in the rib cage (those who make stock from chicken bones will know how easy it is for a human to break the skeleton apart).

Secondly, the sternum (breastbone) is strongly keeled for the attachment of the powerful flight muscles. There are two flight muscles on each side, one of which (pectoralis major) pulls the wing downwards, and the other (supracoracoideus) that pulls the wing upwards. Both muscles lie in the same plane, and supracoracoideus operates via a tendon that runs through a hole formed between the three bones of the shoulder to insert on the upper surface of the humerus or upper arm bone. Alternate contractions of the muscles thus produce flapping, but this only explains part of the mechanism of bird flight and we don't need more detail here.

The third feature of the skeleton that needs to be understood is that the arm bones are reduced in size and they support the wing, which is much longer than the bones that support it, as it is feathered. Not visible in the image are the feathers that make up most of the wing, consisting of primary flight feathers and the secondary feathers that ensure a smooth profile. When viewed in cross-section, the feathered wing has a convex upper profile and a more concave lower surface. Air thus has to travel further over the upper surface of the outstretched wing, and this means lower pressure and therefore lift, something essential in birds when gliding, and also in fast flight, when only the outer section of the wing is used to generate power.

Birds that take off from the ground (like the pheasant) need large beats of the wings to displace enough air to become airborne, and this requires a high energy output. Like all flapping flight, it also requires an abundance of oxygen, and should we look inside the living bird, we would see not only an efficient circulatory system but also the lung capacity boosted by air sacs located in various parts of the body.

Can angels fly?

Let us suppose that angels are physical beings, as shown in so many paintings and that they can fly between Heaven and Earth. In the right-hand panel of *The Wilton Diptych*,[9] we see angels surrounding the Virgin, and their wings seem to be floating above their backs and they may be attached to the shoulder blade or to the robe. Portrayal of the attachment of angels' wings has always been a headache for painters and sculptors, the latter having the greater problem as their work is mainly exhibited in the round. Painters have the opportunity of presenting angels in front view, but, should a rear view be needed for the composition, there is a challenge. Those that suggest that wings may be attached to the robe have the problem of conveying the idea that the cloth is strong enough to support the wing: those that use the alternative approach have the problem of the need to create a hole in the robe through which the wing passes. If one was in a facetious mood, it could be asked how angels dress, or perhaps they never need to? In his painting *Abraham and the three Angels*, Tiepolo creates no confusion and shows a rear view of wings attached at the shoulder, the angels not having robes on the upper part of their body.[10]

Looking at angels as flying physical beings presents many challenges to a biologist. In contrast to birds, angels show no streamlining, and they have arms as well as (bird) wings. One advantage of this arrangement is that dexterity is maintained, so angels do not need either a beak or its equivalent. Both the arms and the wings need to be supported by the skeleton, but how might this be arranged? There is no obvious sign that the skeleton of angels is lightened in the way that the bird skeleton is, and we get no idea of their anatomy. Then, there's the question of musculature, for angels will require enormous flight muscles to lift their heavy bodies from the ground in powered flight. Even if this is possible, there is a further challenge in that the downstroke of an angel's wings will be inhibited by their insertion on the back. This means that only a part of the downstroke is possible before it becomes obstructed by the impact of the wings on the body. Watch a pheasant take off from the ground and it is clear that a long sweep of the wings is necessary to generate enough downforce to lift the bird upwards.

Two mechanisms can be used by large birds that aid the process of take-off. One is to jump powerfully to give the wings a chance to beat in a long arc, as is seen in pheasants, pigeons, and some birds of prey; the other is to utilise lift on the wing by generating speed over the ground (or water) before vigorous flapping lifts the bird clear. Observe swans taking off from water and albatrosses from land and you will see what I mean; the latter being analogous to aircraft accelerating down a runway, as albatrosses need to run for some time before developing critical air speed for take-off. Swans accelerate on water by using their feet as paddles to gain speed.

Perhaps that is how angels take off, or maybe they glide? Gliding utilises the shape of the wing and there are several paintings of angels with outstretched wings that appear to be gliding. Perhaps, the best known of these are the angels painted by Giotto in the Scrovegni Chapel in Padua. Their robes appear to be breaking into flames around the hem, and I have suggested that this may have resulted from Giotto's observations from nature (Wotton 2015). We know that he saw comets, and recorded them, but what did he make of meteors? Could these high-speed objects, trailing flame, be angels coming from Heaven to Earth? We will never know what Giotto thought; but in the pre-Copernican world, it is possible, and what other explanation can be given for the flaming robes of Giotto's angels?

In addition to powered flight and gliding, angels have been shown to hover. This mode of flight requires that the wings are used like rotors, with the body held near vertically: watch a hummingbird closely when it comes to a feeder containing honey and you will see this. Hovering flight requires an abundant energy source, a very light airframe and light body mass, and high wingbeat frequency. Angels, as humans with bird wings, have none of these characteristics.

We must conclude that angels, as depicted by painters and sculptors as physical beings, cannot fly. There *are* those who believe that angels exist in a physical form and there is no reason to suppose that they do not transform in some way, as in transubstantiation. Yet, angels on land are shown to be identical to angels in Heaven and they also appear to be physical entities in the flight between the two.

Some thoughts about angels from an expert

To gain further insight into angels, I turned to the book *Angels: God's Secret Agents* by Billy Graham, and it contains some passages that are worth quoting (Graham 1975):

> It seems that angels have the ability to change their appearance and shuttle in a flash from the capital glory of heaven to earth and back again....Intrinsically, they do not possess physical bodies, although they may take on physical bodies when God appoints them to special tasks. Furthermore, God has given them no ability to reproduce.
>
> *(Graham 1975)*

> ...angels are created spirit beings who can become visible when necessary. They can appear and disappear. They think, feel, will and display emotions... the Bible teaches about them as oracles of God, who give divine or authoritative decisions and bring messages from God to men. To fulfill this function angels have not infrequently assumed visible human form.
>
> *(Graham 1975)*

> ...We must be aware that angels keep in close and vital contact with all that is happening on the earth. Their knowledge of earthly matters exceeds that of men. We must attest to their invisible presence and unceasing labors.
>
> *(Graham 1975)*

> The Bible seems to indicate that angels do not age, and never says that one was sick...The holy angels will never die.
>
> *(Graham 1975)*

> Nothing in Scripture says that angels must eat to stay alive. But the Bible says that on certain occasions angels in human form did indeed eat.
>
> *(Graham 1975)*

This is helpful, as Graham tells us that angels do not need their wings, so discussion of whether they can fly is pointless – so why show them as having wings, something that is perpetuated by the cover illustration of Graham's book in paperback? It is also useful to know that angels do not reproduce, and the cherubs shown in some paintings are not therefore angels in the process of growing up. These cherubs have very small wings indeed, like the wings of a sparrow, and could not possibly fly: nor are they to be confused with the cherubim that we know have four wings. Could they be painted as human babies to remind us of the infant Jesus? If we have a problem with the images of cherubim/cherubs, what about the winged heads that we sometimes see, especially from works of the counter-Reformation? Are these a shortened form of cherubs or, as they can co-occur with cherubs, are they used to represent souls? Clearly, images where bird wings are attached to human bodies are not just confined to conventional angels – the images with which we are very familiar and can all recognise.

Many paintings, like those by Botticini and Murillo (Figures 10.2 and 10.3), add to the confusion, as do the putti of Italian Renaissance paintings, that look identical to religious cherubs. The intentions of putti are quite different though, as they encourage sexual liaison, and are more closely related to Eros (also winged) than to anything that exists in Heaven.

With that point, let's go back to where we started, for, unlike my practical classes in Animal Form and Function, we are not able to observe angels, conduct experiments on them, dissect them, or study their skeletons.

Birds – and wings – as symbols

While eagles and large black birds are no longer images that are widely understood, the use of doves to represent the Holy Spirit (or as a symbol of peace), and angels having bird wings, are universal, with the latter extending to the secular world. Protestantism, in its various forms, saw a reduction in the use of imagery, but in some parts of the Catholic tradition, imagery is so strong that icons, statues, and paintings are themselves worshipped. Mostly, these images are of the Virgin Mary, as she has such an important role of intercession for Catholics, but she is often attended by winged angels.

I have already mentioned the rapid flight between Heaven and Earth that is achieved by angels, but why do we retain the consistent images of angels with bird wings? Is it because we still envy the ability of birds to fly rapidly from one (part of the) world to another, despite our modern inventions?

Notes

1 See, https://commons.wikimedia.org/wiki/File:Christ_Church_Paignton_Eagle_Lectern. jpg and https://www.buildingconservation.com/articles/brass-eagle-lecterns/brass-eagle-lecterns.htm

2 See, Gallen-Kallela *Boy with a Crow*. https://commons.wikimedia.org/wiki/File: Akseli_Gallen-Kallela_-_Boy_with_a_Crow_-_Google_Art_Project.jpg and van Gogh *Wheatfield with Crows*: https://www.vangoghmuseum.nl/en/collection/s0149v1962

3 Piero *The Baptism of Christ*: https://www.nationalgallery.org.uk/paintings/piero-della-francesca-the-baptism-of-christ

4 See, Piero *The Nativity*: https://www.nationalgallery.org.uk/paintings/piero-della-francesca-the-nativity and Rosetti *Ecce Ancilla Domini!*: https://artuk.org/discover/artworks/ecce-ancilla-domini-the-annunciation-117678

5 See, Perugino *Three Panels from an Altarpiece, Certosa*: https://www.nationalgallery.org.uk/paintings/pietro-perugino-the-virgin-and-child-with-an-angel and Guercino *The Dead Christ mourned by Two Angels*: https://www.nationalgallery.org.uk/paintings/guercino-the-dead-christ-mourned-by-two-angels

6 See, Bouguereau *La Vierge aux anges*: https://commons.wikimedia.org/wiki/File:William-Adolphe_Bouguereau_(1825-1905)_-_Song_of_the_Angels_(1881).jpg and Collins *The Fall of Lucifer*. https://artuk.org/discover/artworks/the-fall-of-lucifer-198267

7 See, Gormley *The Angel of the North* https://commons.wikimedia.org/wiki/Category:
   Angel_of_the_North#/media/File:Fly-Angel.jpg
8 See, Pheasant skeleton https://commons.wikimedia.org/wiki/File:Phasianus_colchicus_
   MHNT_Skeleton.jpg
9 See Unknown Artist *The Wilton Diptych* https://www.nationalgallery.org.uk/paintings/
   english-or-french-the-wilton-diptych
10 See, Tiepolo *Abraham and the three Angels* https://www.museodelprado.es/en/the-
   collection/art-work/abraham-and-the-three-angels/e9032f08-e060-48b5-a33b-
   86996227b4b3

References

Graham, Billy. 1975. *Angels: God's Secret Agents*. London: Hodder and Stoughton.
Ferguson, George. 1961. *Signs & Symbols in Christian Art*. Oxford: Oxford University Press.
Martin, Therese. 2001. The Development of Winged Angels in Early Christian Art. *Espacio,
   Tiempo y Forma, Serie VII, Historia del Arte* 14: 11–29.
Wotton, Roger S. 2009. Angels, Putti, Dragons and Fairies: Believing the Impossible.
   Opticon1826. Issue 7. http://dx.dio.org/10.5334/opt.070906
Wotton, Roger S. 2015. Giotto, Angels, and Heaven. Blog Post, https://rwotton.blogspot.
   com/2015/02/giotto-angels-and-heaven.html

11

EARLY MODERN TOUCANS IN SPACE AND IMAGINATION

Alex Lawrence

Introduction

A flash of gold. A glimpse of saffron plumage under the forest canopy. A beautiful but bizarre bird, hybrid and shocking in form, bearing a seemingly impossible beak. This was a sight witnessed by many a traveller to the New World in the sixteenth century, specifically to the Atlantic coastline of what is now Brazil. Europeans would come to know the creature most commonly as the name given by the Indigenous Tupinamba: *Toucan*.

The Toucan was one of the key 'discoveries' made by Europeans in the sixteenth century. Appearing almost immediately in travellers' tales, cosmographies, natural histories, confessional texts, and even in painting and embroidery, it inspired a host of imaginative responses in European cultures as few species of bird had before. Paul Smith (2007, 75–119) has proven how the Toucan troubled contemporary natural–historical discourses, raised critical questions about anatomy, and generated new conceptions of avian form. But the bird reached well beyond the texts of natural historians: it traversed multiple critical territories and cultural media in only a short space of time.

This chapter tells a story of dissemination, of movement through time, and across space (both real and imaginary). One of this section's core contentions is that airborne travel comprises a transgressive response to the physical and social bounds of the man-made world. With all the possibilities of escape, migration, and long-distance movement afforded by flight, birds possess a unique form of agency, a unique way of being in the world.

This essay also aims to re-imagine and broaden our understanding of 'flight'. 'Flight' may be a *human response to avian life*: a flight of the imagination, if you will, which is legible in cultural production. As the Toucan crossed the Atlantic Ocean in the early modern period, it disrupted the way Europeans perceived, and sought to represent, the natural (and human) life in the world around them. It instigated a

DOI: 10.4324/9781003334767-15

set of critical shifts in the way Europeans thought. And was is through these *shifts* – these disruptive interventions in European perception – that we are able to locate a different form of agency in the Toucan.

The research questions we broach in this chapter are broad: what specific impact did the Toucan have on European perceptions of the world? How might we go about locating this impact on the level of cultural production? To put it differently, what kinds of imaginative 'flight' did the bird inspire? What were the lasting effects of these 'flights' in thought and discourse?

The chapter argues that these 'flights' (critical shifts in European thought) occur in *three* principal areas. The first is *epistemic*: so challenging was the Toucan in physical form that it raised a whole gamut of questions about form, anatomy, and the very categories of animal classification. Furthermore, it required Europeans to think differently about how to convey such epistemically shocking information to their uninitiated audiences. The second is *social*. As material samples of the bird (beaks, bones, and feathers) were traded, sold, and gifted in a transatlantic culture of curiosity, it changed the way social actors interacted with one another. As the Toucan (as a material specimen) passed through the hands of explorers, merchants, artists, and nobles, it destabilized, and reconstituted modes of social interaction. The third is *aesthetic*. With its iridescent plumage, multicoloured eye, and bright yellow bill, the Toucan inspired Europeans to reflect on the categories of *taste* and *beauty*, and how to convey aesthetic quality in text and image. A rich current of recent scholarship has unravelled the epistemic and social histories of natural–historical knowledge making in the early modern period (Garrod 2018; Curry et al. 2018; Bleichmar 2012; Daston and Park 1998; Findlen 1994). But only little attention has thus far been given to the aesthetic dimension.

Although we do not have time to discuss every depiction of the Toucan in the sixteenth century (for many such depictions exist), we bring to light a selection of closely related sources. We begin with the *Toucan* of the French chronicler André Thevet, moving onto the *Pica Bressillica* of the natural historians Conrad Gessner and Ulisse Aldrovandi, before ending with the *Byrd of America* stitched by Mary, Queen of Scots. Taken together, these depictions construct a narrative of cultural transmission; they prove the Toucan's unique ability to migrate the real and imaginary spaces of the early modern world.

André Thevet's *Singularitez de la France antarctique* (1557)

In 1555, the Franciscan chronicler André Thevet joined the French Protestant expedition to Brazil led by Nicolas Durand de Villegagnon. The party disembarked on an island in Guanabara Bay (Rio de Janeiro), from whose vantage point Thevet began to describe the human- and natural life of the surrounding region. Although he spent much of his ten-week sojourn laid up with fever, Thevet managed to produce a detailed *relation* of what he saw (Lestringant 2011, 18). He published his account in Paris in 1557, under the title of *Les Singularitez de la France antarctique*. Although the very first mention of the bird in a European text appeared in Gonzalo

FIGURE 11.1 *D'un oyseau nommé Toucan* [Of a bird named *Toucan*] in Thevet's *Singularitez*. BNF/Gallica.

Source: *D'un oyseau nommé Toucan*, Engraving, in André Thevet, *Les Singularitez de la France antarctique* (Paris: Maurice de la Porte, 1557), 91r.

Fernández de Oviedo's (1526, 56) *De la natural hystoria de las Indias*, and the French natural historian Pierre Belon (1555, 184) had commented on a Toucan beak two years prior, the *Singularitez* was the first book to give an eyewitness account of the bird in its native environment. It also included a woodblock engraving (see Figure 11.1). Thevet transcribes the Tupinamba name for the bird – *Toucan* – into text for the first time. The passage is therefore of critical importance to our narrative of transmission: we quote Hacket's 1568 translation in full.

> On the sea coast, the most frequent Marchandise, is the fethers of a birde, that is named in their language *Toucan*, the properties of which I will describe, seeing it commeth to purpose. This birde is of the greatnesse of a Pigeon: there is another kinde like to a Pie, of like fethers that the other have, that is to wit, bothe twaine blacke, saving that about the taile, there are some red fethers among the blacke. Under the brest, the fethers are yellow about foure fingers broade, as well in bredth as in lengthe, and it is not possible to finde yellow more excellent, nor finer in colour than is the fethers of this birde: at the ende of the taile, there are little fethers as redde as bloud. The wilde men take the skin of that parte that is yellow, and they use it to make garnishings of swordes after their manner, and certaine garments, hattes, and other things. I the author of this worke, brought a hatte of fethers very riche and faire out of America, the which was presented to the King of *Fraunce*, *Henry* by name, as a precious jewell. [...] To the rest, this birde is disformed and monsterous, having the bill more greater and more longer than the rest of the body. I have also broughte one of them from thence that was given me, with the skinnes of many of divers coloures: some as redde as fine scarlet, others yelow, blewe, and others of divers coloures.

(Hacket 1568, 73–74)

Reading through the passage, we are encouraged to ask, what kind of an imaginative response did the chronicler have to the bird? To make sense of it, Thevet sets out a series of epistemic claims that bring the bird into the European frame of reference. It is similar in stature to a common European bird – the *Pigeon* – and there is another species similar to a *Pie* (magpie). The comparisons give the reader a sense of scale, but they also, perhaps surprisingly, make the Toucan appear unremarkable. Thevet's description of the beak, however, tells a different story. He is shocked by the anatomical feature, which leads him to describe the bird as *disformed and monsterous* ('difforme et monstrueux'). The term 'monstrueux', as well as outlining a distinct ontological category, has a rich set of cultural meanings in sixteenth-century French writing (Williams 2011; Céard 1977). Thevet's recourse to the term here draws the bird into a wider discourse of monstrosity, then, and proves how the Toucan re-ignited debate about the categories of natural–historical knowledge classification. The chronicler was left pondering: is it a 'bird', an 'animal', or a 'monster'? It does not easily fit within *any* of these categories. Thevet's description thus implicitly raises a critical question about structure: are the existing categories (or existing frameworks) in natural history still fit for purpose?

Furthermore, the French 'difforme' (*disformed*) – which is commonly used to denote the 'monstrous' in literature of the period – promotes the idea that the Toucan is a hybrid creature made of two (uneasily combined) halves. The *body*, on one hand, is small and unremarkable. But the *beak*, on the other, is enormous and so incomparable to anything hitherto witnessed. For the sixteenth-century European, it did not make sense for such a small bird to bear such a large beak. There was a real sense of incongruity in the combination, which sparked the intellectual wonder of the observer and incited the epistemic debate.

The engraving further illustrates the challenge of the beak-to-body ratio. The bird perches in a triangular pose with the beak pointed downwards, which gives the impression that the beak is made of a heavy, dense material, unwieldy to the small anatomical frame. Moreover, the engraver places a nostril halfway along the upper mandible. Both features speak to the way the Toucan forces the natural historian to think differently about anatomical composition in avian species. Never before had Europeans witnessed a beak so large, nor had they seen one without conduits for breathing. The engraver responds to the epistemic challenge by *inventing* – that is, by erroneously placing the nostril and furthering the idea that the beak was too large and heavy for the bird. What the image indicates, then, is the extent to which existing analytical frameworks were no longer applicable. Natural history would need to change dramatically in order to incorporate this hitherto unknown species from the New World.

What makes the engraving more problematic is that the author claims to have a beak of his own. If this were the case and Thevet *knew* what a real beak looked like, why would he depict a weighty, nostril-bearing beak in the engraving? There is a friction between text and image that casts doubt on the authority of the book itself. What we do know is that the engraver and the writer were not the same person (Lestringant 2011, 5). The *Singularitez* therefore presents two competing voices from different actors – one in the text, one in the image. The engraver, for his part, does not accept the testimony of the writer: a bird so small could simply not have a beak so big – the combination is too improbable.

The friction raises questions about editorial process and the location of epistemic authority. Are travellers' reports reliable? Who should we believe? Whose information about the New World can we really trust? Interpretations of the Toucan contrasted with one another significantly, and in the case of the *Singularitez*, diverging views were even visible within the same text. Each individual made their own observation and, ultimately, drew their own conclusion. What is at stake, then, is *rhetorical strategy*, or the act of making one's account more trustworthy than the next before an uninitiated audience. Thevet privileges eyewitness observation. On the narratological level, he positions himself – the *je* ('I') – as the principal agent in the passage: *I the author of this worke*; *I have also broughte*. It was *I*, he claims in continual refrain, who saw the bird, and *I* who brought samples back to Europe.

Thevet develops a distinct rhetorical approach in order to depict the Toucan. For him, it was not enough to simply describe the bird. In order to render his account credible, he frames his presentation in a specific way. In other words, the bird had a disruptive effect on Thevet's rhetorical approach. Faced with the Toucan, the chronicler understood that he must do something differently in order to generate belief: his response – or imaginary 'flight' – was to emphasize his own eyewitness testimony (privileging the *je* ['I']) in the text.

From the passage, we also deduce that the Toucan had a prominent role in a social culture of curiosity in the transatlantic world (Ogilvie 2006, 13–14; Daston and Park 1998, 148–159). Social convention governed the movement of objects between the New and Old Worlds. The nature of trade transactions varied. Specimens were sold to collectors; beaks and bones were shared with natural historians; feathers were gifted to patrons. The influx of *Americana* – Toucan parts in particular – fostered new networks of correspondence and exchange in Europe (Mason 1994, 2–8). Thevet frames his description of the bird within the context of trade. For the Tupinamba, he contends, Toucan feathers have a high social value. The Indigenous social value then bleeds into the European one. Thevet, too, views the samples as *useful* – they contain a distinct social (as well as aesthetic) value and will make a fine gift for his own patron, the French monarch. To give the King such a gift would be to gain favour at court, and, therefore, to advance one's own political standing and ensure continued support for future expeditions (Davies 2000, 137; Daston and Park 1998, 68).

As mentioned, Thevet stresses his role as the key agent in transferring Toucan specimens across space and social strata: *I the author of this worke, brought a hatte of fethers [...] the which was presented to the King of Fraunce*. Thevet constructs his own social authority with the continued use of the first-person singular, and through articulating his proximity to the King. Furthermore, he mentions another set of specimens brought back from the expedition: *I have also broughte one of them from thence that was given me*. His insistence on *moy* ('me') is critical, for, as he claims that samples are given *to him* in the New World, he momentarily occupies the position of the other gift receiver in the passage: the King. The Toucan presents Thevet with the opportunity to further his own social authority. Through the *Singularitez*, we witness a key example of how the Toucan changes (or, has a disruptive impact on) social dynamics in Europe – in this case, the bird presented a unique opportunity for the cosmographer to further his interests at the French Court.

The social value correlates directly with the aesthetic value of the bird's plumage. The coloration of the feathers fascinated Thevet: he describes the tailfeathers as *rouge comme sang* ('redde as bloud') and *rouge comme fine escarlate* ('redde as fine scarlet'). In respect of the breast feathers, it is not possible to find a *jaune plus excellent* ('yellow more excellent, nor finer in colour'). The description shows a clear aesthetic appreciation of Toucan feathers. Indeed, the aesthetic quality also arises from the Indigenous objects fashioned of Toucan feathers: the hat, for example, is *fort beau et riche* ('very riche and faire'), and Thevet gives it to the King not only as a *chose singuliere* – a 'singular thing' or 'curiosity', but also, as we read in Hacket's translation, something that is strongly connected to beauty: *a precious jewell*.

As we can see in the *Singularitez*, Toucan feathers, and their application by Indigenous people, incited two significant developments in European modes of aesthetic perception. Firstly, they increased the limits of colour: never before had the narrator seen such an *excellent* yellow, and the reds were of variegated shading. Secondly, Toucan feathers allowed Thevet to introduce what Stefan Hanß (2019, 587) describes as an 'aesthetics of ingenuity.' All across the Americas, and indeed in Europe, the artisanal dexterity, and technical mastery of Indigenous featherworks, astonished explorers, writers, and artists of different stripes (Rublack 2021, 39). The Tupinamba Toucan-made-objects in the *Singularitez* reflect this Indigenous ingenuity through their unique aesthetic quality. Indeed, in Thevet's (1575, 938r–938v) later account – *La Cosmographie universelle* – he elevates Toucan featherworks even further, describing them as comparable in quality to French silk. The materiality of the Toucan therefore sparked a new form of aesthetic appreciation in Europe, changing the way Europeans thought, not just about Indigenous people and the objects they made – which fed into contemporary discussions about 'savagery' in the New World and the manner in which beauty was perceived. The bird thus helped to open up a new category in early modern aesthetic discourse – one that stemmed from Indigenous ingenuity.

In each of the three areas – epistemic, social, and aesthetic – the Toucan inspired a set of shifts in perception. The bird caused problems, raised questions, and opened new avenues of thought in the *Singularitez* of Thevet. These were the imaginative 'flights' instigated by the bird in the sixteenth-century European observer.

Conrad Gessner *Icones Avium Omnium* (1560)

It was not long before the Toucan appeared again in text. In 1560, the Swiss natural historian Conrad Gessner published a short version of the third book of his *Historia animalium* (1555) – the *Icones Avium Omnium*. The *Icones* is a book of images, a compact edition intended, as Kusukawa (2012, 56) and Smith (2018, 20) argue, to make the expensive woodcut illustrations more economically viable. Yet, these volumes were also vitally important to the overall project: they provided a space in which corrections and revisions could be made (Glardon 2016, 23).

Gessner (1560, 130) depicts a Toucan in the appendix of his third order – 'Avibus volacibus, quae non rapaces sunt, majoribus ac mediis' ('Flying birds, which are not predatory, and of medium or large size'). He ascribes the bird with formal names in classical languages: 'Pica Bressilica' ('Brazilian Magpie or Jay') in Latin, 'Burynchus' or 'Ramphestes' ('beautiful snout' or 'big beak') in Greek, alongside the vernacular forms in German and French. What this profusion of names shows is just how quickly information about the bird had spread since its arrival in Europe. A second definition of 'flight' therefore emerges at this point. From the early accounts of travellers like Oviedo and Thevet, the Toucan – as a *represented* being – took flight beyond its initial (con)text. In only a short space of time, it traversed the borders and languages of Europe and, by 1560, the bird was part of a pan-European natural–historical discourse. The Toucan possessed a unique ability to take flight within the cultural imagination and, as we can see in the nomenclature, each cultural locale in which the bird landed offered a unique interpretation. Gessner's account assiduously compiles all of these terms and offers a brand new image (Figure 11.2).

> The Brazilian Magpie (or Jay): the notable scholar Giovanni Ferrerio gave me its beak which I have depicted here. I added the rest of the body from the description of the *Singularitez de la France Antarctique* by André Thevet, published in French. This very great beak (says Ferrerio) belongs to a certain bird which was brought back from the region of Brazil: the bird is no bigger than our own Magpie (or Jay) (as those who came back to us from these distant lands have stated). [...] I gathered these things from he who afterwards even sent me a sample from the breast, with feathers remarkable for their bright golden or saffron colour (the rest of the body is black, except for the beginning and end of the tail turning red). André Thevet says that the beak is sturdier and longer than the body which we believe more easily, since it is as thin as a membrane, and almost transparent, very light and hollow, and able to hold much air inside; hence its peculiar characteristic of being deprived of airways for smelling. It is of such thinness to allow smells to enter it more easily; thus, if there had been any opening the beak could easily have broken: and for that same reason the beak seems to have been serrated by nature, so that it can cut anything with less force. Or could it even be that the air while travelling around these quasi teeth that prevent the beak from closing completely, finds its way into the throat and windpipe? A bird with a beak of this size could be named 'Burhynchus' or 'Ramphestes' (as of the certain fish). The inhabitants of America call it *Toucan*. See Thevet chapter 47 of the book already cited.
>
> *(Gessner 1560, 130)*

FIGURE 11.2 *Pica Bressillica* by Conrad Gessner.

Source: Getty Research Institute, *Pica Bressillica*, Engraving, in Conrad Gessner, *Icones Avium Omnium* (Zürich: Christoph Froschauer, 1560), 130.

Gessner's depiction relies on two sources. These are, firstly, a beak specimen sent by *Io Ferrerius Pedemontanus* (Giovanni Ferrerio) and, secondly, the textual description of Thevet. Ferrerio was an Italian theologian based in Paris. He acted as one of Gessner's specimen collectors in France, and indeed beyond – most notably in Scotland (Durkan 1980, 349). But where exactly did Ferrerio obtain the beak specimen?

In his later correspondence, Thevet mentions bringing back three beaks from the New World. He describes keeping two in his *cabinet*, which is to say in the royal collection in Paris, and sending one to Gessner: 'Je luy envoyay par mesme moyen un bec de l'oiseau *Tocan* [...] J'en ay deux de reste dans mon Cabinet à Paris' (Thevet 2006, 376) ('I sent him [Gessner] by the same means a beak of the *Toucan* bird [...] I have two remaining in my Cabinet in Paris'). It therefore seems likely that Thevet is the one responsible for Gessner's beak, and plausible that Ferrerio acted as the *passeur*, or courier, between the two. Ferrerio's involvement highlights the increasingly important role played by agents in the transmission of natural–historical information from the New World. Physical samples moved from travellers' chests to collections and, before long, onto the desks of natural historians. It was the agent who facilitated such a movement. Indeed, it seems likely that many objects from the New World passed through these lines of correspondence. Together, a network of agents and collectors was beginning to foster a European culture of natural–historical knowledge exchange that would form the basis of the seventeenth-century Republic of Letters (Ogilvie 2006, 14; Freedberg 2002, 5; Spary 2000, 12; Findlen 1994, 14–16).

Crucially, the movement of the physical sample raises another point about *flight*. The Toucan, as an *object* (a part material specimen), inspired the intellectual wonder of travellers and collectors in Europe. The bizarre anatomical features raised questions and encouraged observers to share information with their learned colleagues. The object then entered into circulation, passing through the hands of multiple individuals. Thus, even as a material specimen, the Toucan took flight into new locations and contexts. Further to reflecting a distinct mobility within the cultural imagination, then, the bird showed a unique ability to travel as a *physical* (albeit deceased) entity in the period.

Returning to the text, Gessner does more than simply cite Thevet. He also responds, in both text and image, to the epistemic questions Thevet raises. Firstly, Gessner remodels the beak entirely. No nostril appears in the engraving; the surface is smooth; the edges of the mandibles are finely toothed; the tip curves slightly downwards. If the phantom nostril of the *Singularitez* posed the question of breathing, Gessner offers a response in the text of his *Icones*. His first suggestion is a claim of porosity: the beak material is so light and thin that air may pass freely through it. The second, phrased as a question further down, is that air can pass through the notches created by the denticulated edges of the beak. At any rate, he asserts, there can be no nostril as this would compromise the overall structural integrity of the beak.

Gessner's claims stem from a close tactile engagement with the material. It is a careful haptic anatomical analysis that allows the natural historian to reframe the debate about epistemic authority. It allows him to claim, implicitly, that *autoptical* (eyewitness) knowledge should no longer be the exclusive purlieu of the New World traveller: the stay-at-home analyst was also someone who saw things closely. Gessner therefore shifts the epistemic framework from New World *autopsia* (eyewitness observation) to Old World *autopsy*. Having read the *Singularitez*, Gessner would have seen that Thevet's image did not match the beak sample he possessed. The incongruity between traveller's report and material object raised the question of trustworthiness and authority, which forces the shift towards a rhetorical privileging of *autopsy*. Gessner's text reveals, then, one of the critical shifts in thought provoked by the Toucan. Upon its arrival on the desk of the natural historian, the bird calls into question the notion of epistemic authority. In other words, having landed in Gessner's study, the Toucan immediately incites an imaginative flight in the mind of the natural historian: one in which the rhetorical technique of presenting new information is, once again, reconsidered.

Yet even after such close and careful probing, Gessner's text still shows a degree of tentativeness. Although he offers suggestions to the breathing/nostril conundrum, he ultimately gives no definitive answer. Rather, he leaves the claims open, framing the latter as a question. Even to one of the foremost natural historians of the time, the Toucan resisted comprehension.

Finally, Gessner's description shows a distinct aesthetic appreciation of the bird. The feathers are *remarkable for their bright golden or saffron colour* and even the *thin, light,* and *hollow* beak exhibits a form of material beauty. A growing discourse of aesthetics was beginning to emerge in natural–historical texts, spurred on by the unique colouration and material composition of the bird. Colour, in particular, was an increasingly important feature in natural–historical description, it seems, and Gessner's Latin expands the colour palate even further (*Aureo* ['bright golden'] and *croceo* ['saffron'] each evoke a specific form of yellow). Furthermore, the engraving leaves considerable blank space where the colourful features are described in the text. We might read this as an invitation to the reader (or owner of the copy-specific edition) to colour in the image themselves, using the textual detail.

Each of these features, in both text and image, reflect a heightened sensitivity to colour. Indeed, they also appear to acknowledge the challenges posed by the printed medium to the depiction of colour. In order to divulge the unique aesthetic dimension of the bird to the reader, an adjustment would have to be made. Natural historians were beginning to realize that painting with words had its limits. As we will see, responses to the aesthetic quality of the Toucan continued to take place in the printed medium – both within and beyond natural-historical writing. It was the Toucan's colourful plumage that inspired Europeans to consider the role of aesthetics in natural–historical description. The feathers themselves, so bright and lustrous as to defy common conceptions of colour, necessitated a response in the way aesthetic quality was conveyed in text. We therefore witness another one of the imaginative flights the bird inspires, this time in the domain of aesthetic assessment.

Ulisse Aldrovandi: Ornithologiae (1599)

The next account in our story is that of Ulisse Aldrovandi, an Italian scholar and one of the most prolific natural historians of the period. Aldrovandi includes a Toucan in his book on birds, *Ornithologiae hoc est de avibus libri decem* (1599). The volume is a compendium packed with engravings and an impressive lexicon of names in classical and vernacular languages. As we gather from Aldrovandi's description, information about the bird had spread widely across Europe by the late sixteenth century. Opening the text with a moderated version of the name given by Gessner (*Picam Bressilicam*), Aldrovandi (1599, 801) states:

> This bird is called *Picam Bressilicam* [Brazilian Magpie (or Jay)], whose image we present here, because if the beak is taken away, it resembles other species of magpie. It can be found in Brazil, Ramphastos [Big beak], and Hipporynchos [Horse snout], or Burynchos [Beautiful snout] are names assigned to it for the size of its beak, from which, being equal to that of *Pelicanum kiranides*, one can call it Ramphones [Beaks]. The Latin name is *Avis piperivora* [Pepper-eating bird] as they feed on pepper, and the German word for it is *Pfeffer voghel* [Pepper bird], and *Pfefferfracsz* [Pepper eater]: Robert Constantin, who mistakenly confers a red colour on the bird, calls it *Pica barbara* [Savage magpie]; the Italians call it *Gaza di Bressilia* [Brazilian Magpie], and the indigenous people of America, the *Toucham*, as André Thevet reports, having travelled in that country and sent letters back from there.

As we can tell, the bird has now crossed great geographic distance and, moreover, reached the imaginations of many scholars, writers, and natural historians of different cultural backgrounds. The profusion of names proves how interpretations of the bird continue to diverge. Some focus on the pepper diet, others on the similarity with magpies, and others still – either by the strangeness of the physical form or by association with the New World – on its 'savage' nature.

In a similar fashion to Thevet, Aldrovandi argues that two species of Toucan are known to Europeans – he is the first to portray both species in image. Yet, the manner in which the natural historian arrives at this conclusion is curious, and it reveals the importance of printing cultures in a shifting epistemic landscape. His first engraving is a clear adaptation of Gessner's *Pica Bressillica*. There are white patches where coloured feathers should be, three forward-pointing toes, and a black spot at the bill tip. The second one reworks the woodcut engraving in Thevet's second major work – *La Cosmographie universelle*. This bird has a compelling eye, a smooth, long beak, and two forward-facing toes.

Thevet did not return to the New World prior to amending his image of the Toucan in the *Cosmographie*. It is therefore likely that he revised his image from the *Singularitez* subsequent to a scholarly debate (with either Pierre Belon or Conrad Gessner). Thevet did not see a 'new' Toucan, then, but corrected his old image after the fact. In this sense, Gessner and Thevet are both discussing the same old bird. But, owing to the differences between the images in the *Icones avium* and the *Cosmographie*, Aldrovandi believes that *two* different birds are at stake. Thus, in his *Ornithologiae*, he prints both, taking them to be two kinds of Toucan.

As well as showing how influential Thevet and Gessner were in discussions about the bird, Aldrovandi's images reveal how the iconography of the Toucan could take flight beyond its original (con)text. As engravings were copied, re-worked, and re-printed (Kusukawa 2012, 64), visual representations embarked on an autonomous 'flight' into successive works. The flights of these images, as we have seen, were key to the process of understanding the bird on the epistemic level: by reproducing both engravings, Aldrovandi argues that two species of bird exist.

Furthermore, the re-production of imagery outlines an active social dimension within natural–historical practice. In some cases, even the physical printing blocks were shared between publishers. For instance, the Toucan in Ambroise Paré's (2019, 2838) anatomical treatise on monsters, *Des Monstres et prodiges* (first printed in 1575), is exactly the same as the one in Thevet's *Cosmographie*. Further into the description, Aldrovandi (1599, 802) claims to have received a beak specimen from a former student, now working in the Low Countries: *D. Nicolaus Espiletus Insulanus*. The networks of correspondence and physical exchange in which the Toucan flies thus continued to expand in the late sixteenth century. Reading the broad nomenclature, we can appreciate just how many sources Aldrovandi drew upon. The *Ornithologiae* therefore reflects, to a large extent, how the Italian natural historian viewed *collecting* as both a material and a textual practice. Just as he gathered physical samples together in his *studiolo* in Bologna, so too did he collect snippets of text and spoken word for his written volumes.

However, all of these sources had to be identified and organized properly. In *Ornithologiae*, Aldrovandi references each source carefully and provides, where possible, the name of the author responsible. As the book proves, a system of citation began to form in humanist natural-historical volumes in the mid- to late sixteenth century and, as it did, it made evident the large social networks of scholars, agents, traders, linguists, and natural historians behind the text. Moreover, scholars actively recognized the benefit of such a system: as well as making a good account of themselves (increasing their own textual authority), they could also lend credence to their non-textual contributors. As Ann Blair (2016, 81) argues, the practice of citation helped to ensure continued collaboration in the future.

The system of citation in Aldrovandi proves how the Toucan took flight, as both a physical specimen and as a representation in text, across social contexts and borders in Europe. Indeed, the *Ornithologiae* locates precisely where, and by what means, these 'flights' take place. As visual representations of the bird move beyond their original contexts, they reveal which sources are considered most 'reliable', incite further scholarly debate and, in the case of Aldrovandi, lead to classificatory confusions. In both the epistemic and social landscapes of natural history, the flight of the Toucan had a uniquely disruptive effect.

Mary Queen of Scots: 'A Byrd of America' (c. 1569–85)

Our final section shifts the focus from continental Europe to Britain and tells a story of flight across cultural media. In the years between 1569 and 1585, the Scottish monarch Mary Stuart was in exile. Imprisoned by her royal cousin, the English queen Elizabeth I, Stuart spent the majority of that period at the residence of George Talbot, Sixth Earl of Shrewsbury. Here, with the help of Talbot's second wife Elizabeth 'Bess' of Hardwick, she set to embroidering a series of tapestries. In her needlework, Stuart depicted a wide range of animals both real and imaginary, from both New and Old Worlds. One of the most visually striking images Stuart stitched was a Toucan (Figure 11.3). As we can see in the inscription, she dubbed it 'A Byrd of America.'

Although the Toucan had now flown beyond the medium of text altogether, it continued to disrupt European discourses in epistemic, aesthetic, and social senses. The clearest epistemic claim (and one that is also aesthetic) is the distribution of colour on the bird's body. The embroidered medium allowed Stuart to depict colour as no printed text has before. The whites and yellows of the breast are clear; the tip of the tail shows a light red (now faded). The visual appearance of the Toucan is much more clearly defined by the embroidery.

Beyond colour, however, Stuart's epistemic argument is not entirely new. The *Byrd* is quite clearly modelled on André Thevet's Toucan in *Singularitez* (Figure 11.1). In fact, as many scholars have noted already, it is a direct copy: the overall shape, the nostril on the beak, and the twig are identical (Mason 2015, 209–214; Bath 2008, 85). Stuart therefore accepts and reproduces the epistemic claims made by Thevet. But again, Stuart's decision to depict *this* Toucan comes down to an editorial and, ultimately, political decision. We know that Stuart drew on a variety of sources to create her embroideries. Certain figures were based on the iconography

FIGURE 11.3 'A Byrd of America' by Mary Stuart.

Source: *A Byrd of America*, Embroidery by Mary Stuart, 'The Marian Hanging', c. 1569–1585, embroidered silk velvet in silks and silver-gilt threat, 2270 × 2940 mm, Victoria and Albert Museum. © Victoria and Albert Museum, London.

of Conrad Gessner and, as Swain's (1973, 107) early work uncovers, she also possessed her own Toucan beak, which was left in a personal collection in Edinburgh. The reason she settled on Thevet for her Toucan, however, may well have been the result of a deep personal connection to the source material. Stuart's epistemic assertions were therefore heavily inflected by the social context.

Stuart had a close connection with France. Promised to marry the Dauphin, she was collected from Dumbarton at the age of five by the same naval officer who would oversee the French Protestant expedition to Brazil – Nicolas Durand de Villegagnon. Stuart then spent her formative years (1548 to 1561) at the French Court. Acquainted with Villegagnon, and likely present at court for the return of Thevet from Brazil in 1556, Stuart would have been well aware of the French activities in the New World. Indeed, as we can see, she was an avid reader of Thevet's *Singularitez*. Michael Bath (2008, 85) posits that Stuart's own beak specimen had been a gift from the French monarch, Henri II, who had likely obtained it from Thevet. This is certainly plausible. However, we should acknowledge that many

Toucan beaks were in circulation in Europe at the time, and we should not forget about Gessner's agent – Giovanni Ferrerio – and his own Scottish connection.

As Durkan (1980, 349) reminds us, Ferrerio had spent considerable time (about 13 broken years) in Scotland, either at court or instructing the Cistertian monks of Kinloss Abbey. Mason (2015, 209) suggests that Ferrerio sent Thevet's Toucan beak to Gessner *from* Scotland. Once again, this is certainly possible. However, how might we explain the fact that a beak *ends up* in Stuart's collection? It is conceivable that, once he finished writing the *Icones*, Gessner handed the beak back to Ferrerio. The latter could then have passed it to Stuart in Scotland or at least to the courtly collection in Edinburgh. Whatever happened, the circulation of the beak through these networks of acquaintanceship proves two points. Firstly, it shows how rapidly along social lines the Toucan moved as a physical specimen. Secondly, it underscores the pivotal role played by the French Court in facilitating these transfers: that is, the dissemination of *Americana* – in both physical and textual formats – across Europe.

As mentioned, the perceptible colour palate of the Toucan expands dramatically in Stuart's visual depiction. The embroidery offers much more colour than the monochromatic engravings of natural–historical texts, and perhaps also more than actual Toucan specimens, which tended to bleach and fade on the long transoceanic voyage home. Belon (1555, 184), for example, discusses the 'whiteness' of his beak specimen. The *Byrd* therefore marks an important shift in terms of articulating the aesthetic dimension of the Toucan to a European audience. Moreover, the tapestry places the bird alongside a series of creatures considered strange, monstrous, or even mythical – a *phenix*, a *unicorn*, a *sea monke*, and a *rhinocerote of the sea*, amongst others. These creatures feature prominently in books on monsters, such as Paré's *Des monstres et prodiges* (1575) and Pierre Boaistuau's *Histoires prodigieuses* (1564). Stuart's tapestry therefore constructs an aesthetics of the monstrous – a category in which the unusual, inexplicable, and epistemically challenging are contained. The Toucan was an integral part of this category: Stuart clearly saw the bird as belonging within the at once epistemic and aesthetic category of the 'monstrous'.

Constructing the Toucan in this way, Stuart proffers the New World, by association, as a place full of such monsters. The epistemically shocking physical appearance of a Brazilian bird led to a much broader characterization of the geographic region (to some degree) as 'monstrous', which has subsequent environmental, climatic, and even racial implications. The Toucan thus incited an imaginative flight in the European monarch and initiated a broader debate about how to conceptualize the New World as a region – a region, it is important to note, the monarch had never seen.

Conclusion: a mobile microcosm

Given the conditions under which Stuart created the tapestries, we might argue that the stitched figure of the Toucan acted as a kind of *lieu de mémoire* or 'memory space'. Imprisoned in England and distanced from her native and adoptive homelands, Stuart practiced needlework as a remedial activity. The figures she crafts came from a vast range of distant lands, time periods, and narratives. Together, they functioned as a means to transcend physical bounds and cross geographic space and time through the imagination. Scholars have argued that the *Wunderkammer* (curiosity cabinet) – a space in which 'curious' objects are assembled from across the globe – works as a 'microcosm', a symbolic space in which the entire world can be viewed in small (Wintroub 2017, 17 and 23; MacDonald 2002, 663). Stuart's tapestries act in a similar way. Her embroidered representations allowed her to experience the world in condensed form, and to be transported to distant lands and times *through* the act of stitching.

Further to allowing the prisoner to imagine the world beyond in a general sense, her embroidered Toucan also provided a specific link to France. Replicating Thevet's engraving with an artisanal skill likely taught to her by the French queen Catherine de' Medici, Stuart embarked on a nostalgic flight of the imagination back to the French Court and the time she had spent there as a child. The Toucan's evocative power, in a figurative sense, lent Stuart wings of her own: the embroidery afforded her the ability to take flight, to escape the confines of her house arrest, and to traverse a mobile microcosm. The bird thus allowed Stuart to recover a form of agency in her exile, however little it might have been.

At the beginning of this essay, we prefigured 'flight' as a human response to avian life, as something birds are capable of doing within the human imagination. We have now seen many instances of 'flight' in relation to the Toucan's reception in Europe. The bird was transferred as a physical specimen across the Atlantic, into the hands of nobles, into collections and cabinets, and into the studies of natural historians. These transferences comprise one 'flight': that is, flight as a physical movement carried out by humans – but, crucially, one which was prompted by the bird itself, by virtue of its evocative power, strangeness and the sense of epistemic wonder it inspires.

The physical 'flights' of Toucan specimens critically informed a second 'flight'. The bird appeared very quickly in a succession of textual and non-textual representations. Chroniclers, natural historians, and embroiderers read, replicated, and amended one another's versions. The Toucan moved rapidly between texts and visual representations: this is our second 'flight' – a flight of the imagination which is legible at the level of cultural production.

As these imaginative 'flights' took place, the Toucan disrupted and re-shaped early modern perceptions of the natural world. The bird forced travellers, scholars, and nobles to think differently in three major ways. It caused a shift in the epistemic landscape. Modes of natural–historical analysis changed to accommodate the bird, and writers reflected carefully on how to convey epistemically challenging information to an audience. In terms of the social dimension, the Toucan's arrival prompted the creation of new social synergies. Travellers passed beaks to natural historians through agents and collectors. Courtiers and nobles gifted beaks to one another. In this way, networks of patronage and diplomacy were upheld and expanded.

In text, natural historians developed a more defined system of citation to recognize the importance of the many actors behind the written work. The sense of a collective knowledge-making practice – or sociality – within natural history began to take shape. On the aesthetic plane, the Toucan's vivid plumage expanded what Europeans knew about colour. It forced natural historians in particular to reconsider their own modes of aesthetic representation in text. On account of its bizarre physical form, it continued to re-constitute the aesthetic and epistemic categories of the monstrous.

The Toucan therefore had a profound influence on European thought through its many flights in the sixteenth century. This influence – the disruptive and, eventually, transformative impact it had on multiple areas of early modern discourse – reflects a different form of non-human agency, one that is grounded in human response. Over the course of this discussion, I hope to have re-shaped what we mean by 'flight' and how we might read non-human agencies in human cultural production – albeit through a very specific lens. Future studies might productively explore the flights of the many other New World birds into the European cultural imagination, not only through the cultural media discussed here but also in performance and pictorial art. At any rate, it is surely an encouraging sign that we may continue to recognize the formative influence of non-human agents in our own cultures, discourses and worldview.

References

Primary

Aldrovandi, Ulisse. 1599. *Ornithologiae hoc est de Avibus Historiae libri xii*. Bologna: Franciscum de Franciscis.

Belon, Pierre. 1555. *Histoire de la Nature des Oyseaux: Avec leurs Descriptions et Naïfs Portraicts*. Paris: Gilles Corrozet.

Boaistuau, Pierre. 1564. *Histoires Prodigieuses*. Paris: Vincent Norment.

Gessner, Conrad. 1555. *Historiae Animalium liber III. Qui est de Avium Natura*. Zürich: Christoph Froschauer.

Gessner, Conrad. 1560. *Icones Avium Omnium*. Zürich: Christoph Froschauer.

Hacket, Thomas. 1568. *The New Founde Worlde, or Antarctike*. London: Henrie Bynneman. https://www.proquest.com/books/new-found-vvorlde-antarctike-wherin-is-contained/docview/2240917684/se-2?accountid=13042

de Oviedo, Gonzalo Fernández. 1526. *De la Natural Hystoria de las Indias*. Toledo: Ramón de Petras. http://bdh.bne.es/bnesearch/detalle/bdh0000050339

Paré, Ambroise. 2019. *Les Œuvres*, edited by Jean Céard, Evelyne Berroit-Salvadore, and Guylaine Pineau. Paris: Classiques Garnier Numérique. doi:10.15122/isbn.978-2-406-09834-8

Thevet, André. 1557. *Les Singularitez de la France Antarctique Autrement Nommée Amerique*. Paris: Maurice de la Porte.

Thevet, André. 1575. *La Cosmographie Universelle*. Paris: Guillaume Chaudière.

Thevet, André. 2006. *Histoire d'André Thevet Angoumousin, Cosmographe du Roy, de deux Voyages Faits aux Indes Australes, et Occidentales*, edited by Jean-Claude Laborie and Frank Lestringant. Geneva: Droz.

Secondary

Bath, Michael. 2008. *Emblems for a Queen: The Needlework of Mary Queen of Scots.* London: Archetype.

Blair, Ann. 2016. "Conrad Gessner's Paratexts." *Gesnerus* 73 (1): 73–122. https://ezproxy-prd.bodleian.ox.ac.uk:2102/10.1163/22977953-07301004

Bleichmar, Daniela. 2012. *Visible Empire: Botanical Expeditions and Visual Culture in the Hispanic Enlightenment.* Chicago: University of Chicago Press.

Céard, Jean. 1977. *La Nature et les Prodiges: L'Insolite au XVIe siècle, en France.* Geneva: Droz.

Curry, Helen, Nicholas Jardine, James Secord, and Emma Spary, eds. 2018. *Worlds of Natural History.* Cambridge: Cambridge University Press. https://ezproxy-prd.bodleian.ox.ac.uk:2102/10.1017/9781108225229

Daston, Lorraine, and Katharine Park. 1998. *Wonders and the Order of Nature, 1150–1750.* New York: Zone Books.

Davies, Natalie Zemon. 2000. *The Gift in Sixteenth-Century France.* Oxford: Oxford University Press.

Durkan, John. 1980. "Giovanni Ferrerio, Gesner and French Affairs." *Bibliothèque d'Humanisme et Renaissance* 42 (2): 349–360. https://www.jstor.org/stable/20676118

Findlen, Paula. 1994. *Possessing Nature: Museums, Collecting, and Scientific Culture in Early Modern Italy.* Berkeley: University of California Press.

Freedberg, David. 2002. *The Eye of the Lynx: Galileo, his Friends, and the Beginnings of Modern Natural History.* Chicago: University of Chicago Press.

Garrod, Raphaële. 2018. "Introduction. Knowledge and Literature: The Natural-Historical Description as Epistemic Genre?" In *Natural History in Early Modern France: The poetics of an epistemic genre*, edited by Raphaële Garrod and Paul J. Smith, 1–17. Leiden: Brill.

Glardon, Philippe. 2016. "Gessner Studies: State of the Research and new Perspectives on 16th-century Studies in Natural History." *Gesnerus* 73 (1): 7–28. doi:10.24894/Gesn-en.2016.7302.

Hanß, Stefan. 2019. "Material Encounters: Knotting Cultures in Early Modern Peru and Spain." *The Historical Journal* 62 (3): 583–615. doi:10.1017/S0018246X18000468.

Kusukawa, Sachiko. 2012. *Picturing the Book of Nature: Image, Text, and Argument in Sixteenth-Century Human Anatomy and Medical Botany.* Chicago: University of Chicago Press.

Lestringant, Frank. 2011. "Introduction." In *Les Singularitez de la France Antarctique (1557)*, edited by Frank Lestringant. Paris: Chandeigne.

MacDonald, Deanna. 2002. "Collecting a New World: The Ethnographic Collections of Margaret of Austria." *Sixteenth Century Journal* 33 (3): 649–663. doi:10.2307/4144018.

Mason, Peter. 1994. "From Presentation to Representation: *Americana* in Europe." *Journal of the History of Collections* 6 (1): 1–20. https://ezproxy-prd.bodleian.ox.ac.uk:2102/10.1093/jhc/6.1.1

Mason, Peter. 2015. "André Thevet, Pierre Belon and "Americana" in the Embroideries of Mary Queen of Scots." *Journal of the Warburg and Courtauld Institutes* 78 (2015): 207–221. https://www.jstor.org/stable/26321954

Ogilvie, Brian. 2006. *The Science of Describing: Natural history in Renaissance Europe.* Chicago: Chicago University Press.

Rublack, Ulinka. 2021. "Befeathering the European: The Matter of Feathers in the Material Renaissance." *American Historical Group* 126 (1): 19–53. doi:10.1093/ahr/rhab006.

Spary, Emma. 2000. *Utopia's Garden: French Natural History from Old Regime to Revolution.* Chicago: University of Chicago Press.

Smith, Paul J. 2007. "On Toucans and Hornbills: Readings in Early Modern Ornithology from Belon to Buffon." In *Early Modern Zoology: The Construction of Animals in Science, Literature and the Visual Arts,* edited by Karl A. E. Enenkel and Paul J. Smith, 75–119. Leiden: Brill.

Smith, Paul J. 2018. "Deux Recueils d'Illustrations Ornithologiques: Des *Icones Avium* (1555–1560) de Conrad Gessner et les *Portraits d'oyseaux* (1557) de Pierre Belon." In *Natural History in Early Modern France: The Poetics of an epistemic genre,* edited by Raphaële Garrod and Paul J. Smith, 18–45. Leiden: Brill.

Swain, Margaret. 1973. *The Needlework of Mary, Queen of Scots.* New York: Van Nostrand Reinhold.

Williams, Wes. 2011. *Monsters and Their Meanings: Mighty Magic.* Oxford: Oxford University Press.

Wintroub, Michael. 2017. *The Voyage of Thought: Navigating Knowledge Across the Sixteenth Century World.* Cambridge: Cambridge University Press.

12

PEREGRINE FLIGHTS

The emergence of digital winged geographies

William M. Adams, Adam Searle, and Jonathon Turnbull

Introduction

The peregrine falcon (*Falco peregrinus*) is a cosmopolitan bird with a global distribution (Drewitt 2014; Gainzarain et al. 2000; Molard et al. 2007; Ratcliffe 1993). No less than 75 subspecies have been described, although currently only 19 are recognised by biologists (White et al. 2013). Today, the peregrine is not considered a globally endangered species. On the contrary, its population is slowly growing (Birdlife International 2022). Yet, persecution and habitat destruction over the last two centuries (related to intensified agriculture in countries like the UK) resulted in its population falling to such a low level that peregrines seemed on the verge of extinction. Indeed, the peregrine became a poster child of the conservation movement.

The peregrine is widely considered to be highly charismatic (see Lorimer 2007). It is reputed to be the fastest animal on Earth when it stoops to hunt prey, a factoid feted from pub quiz trivia to children's encyclopaedia. In the book *Falcon*, Helen Macdonald describes the long history of human fascination with the peregrine in falconry, for both its elegance as a flyer and its prowess as a killer (Macdonald 2006).

Naturalists have also long singled out the peregrine for its speed, fierceness, and relative unknowability (Ratcliffe 1980). Ed Drewitt describes the peregrine stoop as "mind-boggling" (2014, p. 21), writing that "when one is in stoop dive it drops through the air with speed, grace, and perfection" (2014, p. 20). In the classic work of nature writing *The Peregrine*, J.A. Baker (1967) describes with astonishing eloquence his fleeting encounters with peregrines, which, for him, magically dominate the skies of Essex (UK), picking off their chosen target from the sky with purpose and ease. Flight, Macdonald (2006, p. 25) suggests, is "the single most celebrated falcon characteristic."

DOI: 10.4324/9781003334767-16

Peregrines epitomise that quality quintessential to all flying birds, of being able to escape the lived spaces of human beings to the skies. Humans observe them, admire them, and are awed by them. They intrigue the human imagination, which is driven by their relative elusiveness. Even the falconer's controlling arts place incomplete constraints on the peregrine, and wild birds are uncontained, exuberant, and ephemeral. Flight and predation lie at the core of the peregrine's lived world, and they escape from the flickering and static human gaze with consummate ease. Helen Macdonald (2006, p. 7) writes, peregrines "excite us, seem superior to other birds and exude a dangerous, edgy, natural sublimity."

Peregrines are also significant animals for scholars interested in the relations between humans and living others. They show what is at stake in those relations, especially in the profoundly transformed ecologies of the Anthropocene world. Their history in the long twentieth century was at first tragic and then later transformative. It is impossible to consider peregrines, to represent or to imagine them, without reference to their contacts with humans. Peregrine and human worlds are deeply intertwined.

The physical flight of the peregrine is a natural marvel and a metaphor for the power and strangeness of nature (Macfarlane 2017). But its flight away from, and later back into, the human sphere is equally remarkable. The peregrine's changing lifeways and geographies emerged alongside anthropogenically changed environments. Indeed, they almost faded to extinction, but came back. They were forced to flee human society, but re-joined it, on new terms.

Following one of the book's core provocations, we reflect the language of Gilles Deleuze and Félix Guattari (1987) and explore the "lines of flight" or "lines of escape" [*lignes de fuite*, Massumi 1987, p. xvii] taken by the peregrine, as it slipped away from the human gaze and human landscapes and then returned. We focus on the fact that, in returning, the peregrine has changed. The spaces in which it dwells, its relations and co-dependencies with humans, the epistemological practices of human knowing and knowledge production relating to it, and indeed its very ontological constitution have all undergone fundamental transformation. To follow the peregrine's *lignes de fuite* is to follow the production of multiplicities and changes in nature. Playing with Deleuze and Guattari's *double entendre*, we think through these changes in terms of flight and escape. The peregrine escapes human observation and knowledge practices, transforming in nature, in value, and in meaning. In tracing these moves, we follow other researchers interested in human–avian relations at the interface of the natural and social sciences (e.g., Despret 2021; Garlick 2019; Van Dooren 2014).

In this chapter, we explore three specific peregrine flights. First, we consider the peregrine as a revenant ghost: its transformation into spectral form, its fading from human sight as it neared extinction, and its resurgence and ecological recovery. Second, we consider the peregrine's movement into new spaces; specifically, the built environments of urban landscapes. The revenant peregrine has become urban, occupying novel territories and displaying new mobilities closely aligned to human worlds. Third, we consider a further transition that has accompanied the peregrine's recovery and urban flyways: its emergence in and through the fractal spaces of the digital world. Cameras mounted on nests, shared through the internet, have allowed the peregrine to move through new digital skies, interweaving its flight across a different cloud. Furthermore, it has brought peregrines to the attention of new and broader human publics, transforming not only the way humans perceive, understand, and respond to them but also the dynamics and prospects of their own physical lives through the novel biopolitical regimes digitisation engenders.

Our empirical story addresses ideas of intimacy and proximity in human–avian encounters. Contrary to the notion that these encounters must be haptic and corporeal to be considered genuine, we explore the authenticity of encounters facilitated through digital technologies and mass media, examining the consequences of such interventions for contemporary environmentalism.

The spectral peregrine

Peregrines are a species unavoidably associated with loss, absence, and extinction: key frameworks used to make sense of nature in contemporary environmentalism and environmental scholarship (Searle 2020; Yusoff 2012). Indeed, extinction has long exerted great emotional power among lovers of nature and conservationists (Adams 2004; Barrow 2009; Heise 2016). The peregrine's decline from a common bird to the edge of complete disappearance across Europe and North America narrates a familiar tale of the ecological consequences of industrialised capitalism and agriculture.

Extinction, disappearance, and loss have a spectral quality. Species in decline become hard to observe, disappear, or go extinct and, as such, exist as ghosts in the landscape. Theirs is a liminal existence from a human's perspective; their presence a matter of speculation, of scientific expeditions, of doubt, and haunting (McCorristine and Adams 2020; Searle 2021; Wrigley 2020). On the other hand, species that are rediscovered, re-appear, or start to recover their populations flicker into view like revenants; their presence fragile, temporary, or perhaps even doubted. But revenants don't just *return from* the past. They also *arrive to* altered cultural, political, and ecological contexts. As such, when species rebound, like revenants, their renewed presence renders them uncannily different (Searle 2022). In this light, the peregrine falcon is a classic revenant or "ghost species" (McCorristine and Adams 2020). Its persecution and decline meant their presence haunted former territories; their cliff eyries fell silent, and the skies where it once flew and hunted lay empty and tame. Its recovery, however, involved its return as a revenant, transformed.

Like other birds of prey, peregrines were widely killed because they were predators of gamebirds raised for aristocratic sport shooting. This persecution became heavy in the nineteenth century (Ratcliffe 1980). Gamebirds were shot in small numbers from the sixteenth century, but the invention of the percussion cap gun, and later the breech-loading shotgun, allowed bigger game bags, "driven" game shooting, and specialised shooting estates. The 1831 Game Act allowed non-landowners to kill game, and shooting expanded, receiving Royal endorsement as a proper aristo-cratic pursuit following Queen Victoria's marriage in the 1840s (MacKenzie 1988).

In the nineteenth and early twentieth centuries, peregrine populations suffered from two further threats. The first was the attention of collectors seeking taxi-dermy specimens and egg-collectors (Ratcliffe 1993). Their growing scarcity fur-ther spurred desires to acquire such specimens (see Searle 2021). The second was the killing of birds and destruction of eyries by pigeon racers. In 1925, the National Homing Pigeon Society issued a circular appealing for the removal of all protection for the peregrine, and indeed, calling for its extermination. The RSPB opposed the suggestion, and the Home Office asked the Wild Birds Advisory Committee to consider the issue. Like masterly fence-sitters, they decided the peregrine should continue to be protected, except in areas where there was positive proof of its caus-ing serious losses among homing pigeons (Sheail 1976). Peregrines remained perse-cuted. During the Second World War, the British Air Ministry saw them as a threat to carrier pigeons carrying messages from occupied Europe and "outlawed" them: (Ratcliffe 1963, p. 64) the pairs nesting on the white cliffs of Dover were shot in the interests of national security. Even today, peregrines remain the targets of illegal killing, albeit on a much smaller scale, by those such as pigeon racers outraged by their predatory activities (Humphreys et al. 2007; Ratcliffe 1993).

After the Second World War, the already reduced British peregrine population was driven down further by the adoption of organochlorine pesticides in agricul-ture (Ratcliffe 1970). In *The Peregrine*, published at the height of the pesticide crisis, J.A. Baker (1967, p. 32) wrote apocalyptically of the bird's future in the British Isles: "few peregrines are left, there will be fewer, they may not survive. Many die on their backs, clutching insanely at the sky in their last convulsions, withered and burnt away by the filthy, insidious pollen of farm chemicals." By the 1960s, the UK peregrine population had fallen by 80% compared to its pre-war abundance (RSPB 2021). Empty eyries and broken eggshells came to define the raptor. Tim Dee (2009, p. 89) comments that he "grew up thinking of peregrines as sickly." They had been "ruined by the malign human idiocies that were contaminating the whole world" (2009, p. 89).

Rachel Carson's *Silent Spring* was published in 1962, but it took more than a decade for research to prove both the impact of synthetic pesticides on predatory birds, and how their destructive effect worked. Insecticidal seed treatments caused mortality in wild birds, but it was the persistence of organochlorines in the bodies of predators that hit bird of prey populations. These synthetic chemicals were not broken down in the body and accumulated in the livers of predators at the top of the food chain when they ate fish, mammals, or other birds that had eaten contami-nated seeds and crops. In birds, eggs were laid with thin shells that broke in the nest and sometimes with no shell at all. The breeding success of peregrines (and other raptors) plummeted in the UK and many other countries, and so did their numbers (Greenwood 2021; Newton 2015; Ratcliffe 1963).

Eventually, public campaigning (particularly in the USA and Europe), building on a mass of biological research on the impacts of organochlorine pesticides, led to progressive bans on their use in agriculture in the 1970s and their replacement with more toxic (but less persistent) organophosphorus compounds (Mellanby 1967; Ratcliffe 1963, 1970; Sheail 1985). Populations of peregrines and other raptors began to recover slowly across North America and Europe (Ratcliffe 1984, 1993; Tucker 1998). By 1991, there were 1283 breeding pairs in the UK (Crick and Ratcliffe 1995). As such, the peregrine became a *"cause célèbre* for nature conservation" throughout the second half of the twentieth century (Ratcliffe 1972, p. 118).

For conservationists, more than any other bird, the decline of the peregrine and its brush with extinction spoke for the broader, ongoing impacts of industrial society across the capitalist and industrialised agricultural expanses of Europe and North America. Moreover, its recovery also represented more than simply the return of a single species. Its return to British skies and to cities elsewhere in the world epitomised the hopeful prospect of ecological recovery, the beginning of a new, more rational, and more benign relationship between nature and industrial society. The peregrine was back, a revenant.

The urban peregrine

Yet, whilst ecologists speak of the peregrine's "comeback," the revenant peregrine was not identical to the bird that was lost. Biotic absences can produce strong affective atmospheres of longing and belonging for the departed, as cultural and environmental geographical scholarship on spectral ecologies shows (Garlick 2018; Jørgensen 2019; McCorristine and Adams 2020). However, when ghosts return, their changed habitats, relations, and representations render them uncannily different (Searle 2022). The peregrine's transformation into a ghost, and later a revenant, changed it in several key respects.

First, the recovering peregrine population was, to some extent, genetically different. The growing population contained feral birds that had escaped from captivity, primarily from falconers. Captive breeding and release programmes also sometimes used feral birds, or birds from different populations, thus fundamentally recalibrating what is meant by an authentic peregrine in an age of mass extinction (Schroer 2021). Second, the returned peregrine behaved differently and lived in different territories and landscapes. As populations recovered, birds started to re-occupy historic nest sites on remote sea and inland cliffs. However, the largest growth in breeding numbers came in anthropogenic environments, on urban buildings such as skyscrapers and cathedral spires across the British Isles (Banks et al. 2010; Dixon 2000). Third, in this new urban milieu, the peregrine found new prey and new ways to hunt. It was still a devastatingly effective aerial killer, but it adapted its hunting practice to the new, human-made, urban environment.

The peregrine came back, but not as a bird of remote cliffs and mountains. Instead, this new peregrine was a bird of the city: skyscrapers, cathedrals, and factory chimneys mimicked the natural vantage points that cliff provided peregrines for hunting and also functioned as nesting sites (Drewitt 2014). They provided the scaffolding around which new lives and new ecologies were woven. Yet, importantly, this shift was a powerful cultural symbol as well as a biological reality. The peregrine's move into built spaces contributed to, and for many observers came to represent, a wider awareness of the diversity of urban nature and the opportunities for its conservation, and the peregrine began to inspire hope for conservation success in cities globally (Macfarlane 2017).

Wild species of all kinds were found lodging within urban topographies and material infrastructures built for human purposes. As recombinant and cosmopolitan ecologies emerged (Hinchliffe and Whatmore 2006; Hinchliffe et al. 2005; Lorimer 2008a), human conceptions of nature expanded and hybridised to reflect new visions of collaborative futures (e.g., Marris 2010; Schilthuizen 2018). As Maan Barua (2022) writes regarding the complex multispecies stories of peregrines and feral ring-necked parakeets in London, ecological recombinance opens up nonbinary conceptions of what living beings should or should not belong in the city and generates a range of ethical and affective responses. With this increased awareness and heightened public sentiment towards the urban peregrine, came new organisations: in the UK, the Urban Wildlife group was established in Birmingham in 1980 and the London Wildlife Trust in 1981 (Goode 2014).

The peregrine, with its wildness and recovery story, took pride of place among the new urban fauna. The peregrine's transition "caught the imagination of many of the human city dwellers with whom they now share territories" (Hennen and Macnamara 2017, p. xiii). The peregrine became a conservation "flagship species" (Donázar et al. 2016), in part because of its ability to thrive in the interstices of urban spaces. Previous epistemological frames built around the peregrine as a bird of remote places, hard to see and in drastic population decline, were overwritten by their presence and visibility in urban space. They became a familiar sight in urban skies, woven into the quotidian fabric of human urban life. As Chris Packham writes (2014, p. ix) in the forward to Drewitt's *Urban Peregrines*: "on the rainy streets of Norwich, the sunny sidewalks of Chichester or the bustling pavements of Bristol, we can stop shopping and watch the best avian royalty reality show on earth."

The peregrine's flight into new urban habitats involved changes in ecology and behaviour. Cities provided rich opportunities for peregrines to hunt and feed (Drewitt and Dixon 2008; Kettel et al. 2019; Mak et al. 2021a, 2021b). Tall buildings juxtaposed with low-lying spaces such as parks or industrial sites resembled cliff sites which offered "commanding views of the surrounding landscape" (Mak et al. 2021a, p. 7). Height allows peregrines to observe their surroundings and provides time to increase stoop speed and consequential predation success (Drewitt 2014; Jenkins 2000; Ratcliffe 1993; Time 2016).

Artificial light also provides new opportunities for hunting (Kettel et al. 2016; Stirling-Aird 2015). Feral pigeons are ubiquitous and a favourite prey. But urban peregrines have been shown to capture a large diversity of other birds, often those moving above the city on migration routes: woodcock for example, or curlews, or ducks. Some birds have been shown to use the illumination of city lights to allow nocturnal hunting, something not previously recorded (Kettel et al. 2016).

City hunting is so rich that urban peregrines need smaller territories. Cities are therefore able to support higher population densities than other habitats.

At the same time, of course, the designation of "the city" (however beloved by geographers and other humans) makes little sense to a hunting peregrine. From many cities where peregrines now nest on prominent buildings (in Sheffield, e.g., or Wakefield, or Exeter in the UK), it is but a short flight to open country. Urban Salisbury – whimsically regarded a city due to its cathedral, on which a celebrated peregrine pair breeds – is not ecologically comparable to the sprawling metropolitan area that envelopes nesting sites in central London. In many cases, the revenant peregrine has a perch in the city, but its hunting eyes are focused far afield.

Peregrines certainly now find cities relatively safe places to nest and rear their young. Conflicts with humans are few in relation to historical rural frictions (Pagel et al. 2018; Washburn 2018; Wilson et al. 2018). They are in the public eye, and many, as we describe below, are closely watched over and monitored. Injured peregrines have sometimes been rescued and nursed back to health, and people who might wish to damage a nest or molest a bird are discouraged by the surveillance infrastructures and business of the city. It is not easy to scale a cathedral tower unobserved, as it might be a sea cliff. Contrary to common representations of urban areas as sites of antibiotic control unsuited to wildlife (Lorimer 2020), for the peregrine, the city is a sanctuary.

The digital peregrine

As revenant peregrines began to populate the clouds above city streets, they also began to take wing in a different type of cloud, one composed of electrons, digital code, and computer software. This was perhaps their greatest transformation and had profound implications for environmental governance and wildlife conservation. By coming into cities to breed at the time they did, they entered into a world now saturated with digital data, computer visualisation, and the internet. The revenant peregrine became not only urban but also digital.

New technological entanglements, in recombinant urban ecologies, enable new regimes of visibility and management. Just as urban infrastructures enabled successful peregrine adaptation due to their material topographies, abundant ecologies, and sanctuary from persecution, they have also facilitated greater mediation and dissemination through the widespread presence of information communication technologies. Indeed, livestreamed audiovisual footage of peregrine nests in urban sites around the world is now commonplace.

The critical technology with which the peregrine became entangled was the webcam, or "nestcam." The first webcam was invented in 1991 and broadcast via a local network. The first online streaming began in 1993 (Campanella 2004). In the early days of this networked technology, webcams transmitted simple "JPEG refresh" footage: low-quality static images that refreshed at regular intervals. Due to this limitation, webcams in the mid-1990s focused on mundane happenings such as weather updates. Yet, a number of keen birders, who had been eagerly observing the peregrine's urban adaptation, began reimagining these technologies as a site of encounter with, and mediation of, a bird that had for decades eluded their gazes. The first peregrine nestcam in the world gazed into a nest on the Sun Life Financial Centre in Toronto, Canada, in 1997 (Drewitt 2014). The second was in the UK, installed on the roof of the Sussex Heights building, a 24-storey residential tower block on the seafront in Brighton, in March 1998 (Drewitt 2014). Live images of a peregrine falcon were transmitted over the internet for the first time in Europe. A global audience began to grow.

By the turn of the millennium, surveillance cameras had been widely installed across British cities as part of an expanding urban panopticon (Koskela 2003), and their application to wildlife observation also expanded. CCTV technologies provided uninterrupted analogue feed, which could be simultaneously digitised and broadcast across the internet. Internet protocol (IP) camera systems, appearing in the mid-2010s, made simultaneous livestreaming easier, faster, and cheaper. The capacity for livestreaming directly to online platforms such as YouTube emerged in the late-2010s. This series of successive technological innovations altered the ways that the peregrine was captured through digital mediation. This included the introduction of high-definition cameras, controllable "pan, tilt, zoom" systems, and the addition of audio streams, and infrared viewing allowed observers to see and experience more of peregrine life and ecology (Figure 12.1).

The peregrine's urban resurgence is just one part of a wider digital emergence of urban natures (Moss et al. 2021) and parallels the development of digital technologies in conservation more generally (Adams 2019; Arts et al. 2015; Chambers 2007;

FIGURE 12.1 Peregrine falcon above the Sheffield skyline, as broadcast to the world one morning in February 2020. Image reproduced courtesy of Sheffield Peregrines. http://peregrine.group.shef.ac.uk/.

Scoville et al. 2021; van der Wal and Arts 2015; Verma et al. 2015). Surveillance technologies have been central to the deeper entanglement of digital media and non-human life, which have been developed in a range of settings (von Essen et al. 2021).

The digital peregrine emerges from livestreamed images captured by cameras mounted high above city streets, on or near nests. Installation of cameras, and the simultaneous streaming of their images, altered the way peregrines were seen and understood. The impact produced new epistemic practices and ethological knowledges that changed what was known about peregrines, whilst also creating a series of digital spaces for interpersonal and affective engagements between observing publics and perhaps unknowing peregrines. Digital media allowed behaviour to be observed that had never previously been observed (the identification of food items, e.g., or the way chicks and parents interacted). They also made it possible for people to see these things who could never previously have done so, even at a distance. It was no longer necessary to perch on a hide halfway up a sea cliff to watch a peregrine at home, or even to fill in forms applying for a licence to do so. Moreover, the expanded sensing capabilities offered through technological innovation allowed for novel insights into their ecology; for example, infrared cameras allowed Kettel et al. (2016) to observe how urban peregrines hunt and feed their young at night.

Digitisation brings both naturalists and members of the public closer to peregrines. Many nestcam streaming services now offer viewers the ability to switch between various nestcams aimed at the same nestbox, hinting towards a feeling of "remote control" (Oliver 2021). Usually, one of the nestcams will provide an overhead perspective, allowing close inspection, whilst additional nestcams offer views of peregrines in their airy urban settings. Nestcams share characteristics with "slow television" (Jørgensen 2014; von Essen 2021). People watching nestcams can form close attachments with individual birds, watching unfiltered streams of their lives from birth to fledging. Watchers are in part attracted by the calmness and predictability of peregrine livestreams, as well as the opportunity to attune to other animals' atmospheres (see Lorimer 2008b; Lorimer et al. 2019). Drewitt (2014, p. 118) suggests that livestreaming of the peregrine nest on Sussex Heights was "the very beginning of many people's enjoyment, fascination, and hobby (even bordering on obsession!) of watching peregrines online from nests around the globe."

The engagement of watchers with livestreamed peregrine webcams can involve emotional bonds. The first fledging of the 2021 season was described by the Chichester Peregrines blog as "a very emotional moment," a sentiment echoed on social media by people watching peregrine nestcams around the world (Chichester Peregrines 2021, n.p.). Such emotional engagements can span impressive distances: one visitor arrived at Norwich Cathedral's peregrine viewing point having travelled from Tasmania because they had been following their lives online. When the peregrine nestcam at Sussex Heights in Brighton went live in 1999, its first audience was predominantly in Germany. Nestcams had been adopted in Germany earlier than in the UK, and there was a substantial audience of German livestream observers when the Sussex Heights camera went online.

The deep affection felt for particular peregrines, and their broadcast in the mundane spaces of daily human life, such as homes and offices, can lead to forms of "digital domestication" as observers project forms of anthropomorphic personality onto the animals (Kamphof 2011; Searle et al. 2023). In addition to moments of wonder and intrigue, nestcams also portray the visceral and bloody side of peregrine ecology, which comes with some backlash (Pitas 2022). The killing of prey and competition among broods of chicks – peregrines start incubating as soon as the first egg is laid, so there is always an oldest chick and a runt – create moments of drama and emotion, blood and mortality that often challenge or shock observers.

Of course, digital peregrines did not only emerge through nestcams and livestreaming. Digital animals are multiple, existing as streams of data and in server farms around the world (Adams 2020). Ecologists in Italy, for example, attached GPS trackers and video cameras to juvenile peregrines to study their hunting (Brighton et al. 2017). Nestcams, however, remain the predominant variant of the digital peregrine and the most publicly accessible and widely encountered. The digital peregrine is the outcome of a complex process of what Sheila Jasanoff (2004, p. 15) calls "constitutive co-production,", involving the "emergence of new socio-technical formations," between people, birds, and technologies.

Encounters with digital peregrines provoke affects that are distinct from those produced through encounters with actual peregrines. The up-close-and-personal view afforded by nestcams provides access to intimate moments in peregrines' lives that emphasise cuteness and vulnerability over fierceness and magnificence, which are more commonly associated with actual peregrine encounters. This is because certain aspects of peregrine lives, such as their famous hunting stoop, are not visible via nestcams. In addition, it is not uncommon for webcam viewers to experience awkwardness related to a sense that they are invading the private lives of peregrines.

In this light, the digital peregrine can be understood as an assemblage of corporeality, data, and visualisation practices. Webcams bridge the physical and cybernetic, projecting and mingling the images and agencies of both (Campanella 2004; Stefik 1997). The "body individuals" they capture "become, in one sense, intertwined with digital individuals" (Koskela 2004, p. 200). There is therefore no clear distinction between the "real/actual" and "digital/virtual" peregrine (see Ash et al. 2018), especially when viewers are watching live. Digital animals in mediated networks complicate the binary between digital and physical space (Berland 2019; Stinson 2017). Following McLean (2020), the digital peregrine is not just a representation of an actual peregrine. It can be thought of as "more-than-real" (McLean 2020), with a transitory status in translation between physical and cybernetic forms (Rose 2015). Although partial, the digital peregrine is not less-than-actual, a diminished form of a "real" peregrine, it is more-than-real, an encounterable entity in its own right, although moored to the actual corporeal peregrine.

Encounters with digital peregrines, brought into being through livestreaming, offer specific opportunities for the creation of meaningful emotional connections with human watchers. These are related to the affects produced by direct physical observation but can both be experienced by people who are not (and perhaps can never be) physically present to observe the bird and can reveal sights or behaviours that no ordinary human observer could see (close images of the occupied nest for example). The livestreamed digital peregrines play a key role in the popularity of the species, and the potential for convivial human–bird relations, albeit mediated through the digital realm. Unlike the spectacular, disengaged encounters that take place in wildlife documentaries and pre-recorded YouTube videos, nestcams involve viewers in the immediate lives of peregrines, engendering a palpable sense that one is witnessing a living, breathing creature.

Encounters with digital peregrines sometimes directly affect actual corporeal peregrines. Livestreamed nestcams broadcast the locations of peregrine nests to the public, allowing for surveillance to take place around the clock as part of an emerging urban polyopticon (Allen 1994). As a result, urban peregrine nesting sites tend to be considerably safer from vandals or persecutors than remote locations, not only due to their physical inaccessibility but also the ubiquity of their monitoring. Urban environments therefore emerge as spaces of refuge for peregrines, aided by the social networks of their digital guardians. The life-chances of the living peregrine are, in part, aided by its digital avatar.

In observing livestreamed peregrines online, webcam watchers grasp only a limited version of the realities of free-flying physical birds. For a few weeks, juveniles continue to take small flights, increasing in frequency and duration before eventually leaving the nestbox. During this time, they come and go on viewers' screens, but afterwards, they disappear. Peregrine lives thus partly evade digitisation. Every time the peregrine moves off the nestbox or perch in search of prey or to stretch its wings, it escapes the nestcam's panoptic gaze. As such, the digital peregrine goes feral as its intractable wildness eludes the entanglement of digital media. The fact that the digital peregrine is only part of a wider story is integral to its charisma (see Lorimer 2007). Digitisation signals one line of flight that has seen the transformation of the peregrine. It can only be understood in relation to the other multiplicities of relations emerging through spectrality and urbanisation.

Conclusions and future directions

In our exploration of the recent natural, cultural, and technological history of the peregrine falcon, three specific lines of flight offer provocation for scholars addressing the interface of the natural and social sciences and the overlaps of human and avian worlds. The peregrine's multiple and complex peregrinations through time and space have taken in near extinction as a spectral species, re-emergence as an opportunist urbanite, and a translation into digital existence as a cosmopolitan animal enjoyed by people around the world. These movements speak to broader dimensions of the complex relationships between humans and other animals that are characteristic of the contemporary ecological condition.

To explore these *lignes de fuite*, we need to follow the peregrine through different paradigms, to understand the elusive moments of transformation. The idea of spectrality frames the peregrine's line of flight through biotic decline and recovery, and by implication absence and presence. The idea of ecological recombination describes the character of the new urban habitat that the peregrine adopted. The urban peregrine moves seamlessly between zones classified by human observers as rural and urban. It went from being a relatively invisible animal to one whose encounters have become an integral element of the everyday urban fabric. In exploring its digitisation, the peregrine became unmoored from merely physical skies and roofs, and began to move through global digital spaces, claiming attention from human observers around the world. Indeed, as the story of these birds shows, peregrines have come to epitomise animal cosmopolitanism in the twenty-first century.

Acknowledgements

We would like to thank the editors of this book for their advice and feedback on our chapter, in addition to the attendees of the Winged Geographies research seminar who provided stimulating points for discussion. Without the insight and openness of ecologists, conservationists, and nestcam hosts engaged in peregrine falcon conservation in the UK, this chapter would not have been possible.

References

Adams, William M. 2004. *Against Extinction: The Story of Conservation*. London: Earthscan.

Adams, William M. 2019. "Geographies of Conservation II: Technology, Surveillance and Conservation by Algorithm." *Progress in Human Geography*, 43: 337–350. https://doi.org/10.1177/0309132517740220

Adams, William M. 2020. "Digital Animals." *The Philosopher* 108: 17–21.

Allen, M. 1994. 'See You in the City!' Perth's Citiplace and the Space of Surveillance." In *Metropolis Now: Planning and the Urban in Contemporary Australia*, edited by Katherine Gibson and Sophie Watson, 137–147. Melbourne: Pluto Press.

Arts, Koen, René van der Wal, and William M. Adams. 2015. "Digital Technology and the Conservation of Nature." *Ambio* 44: 661–673. https://doi.org/10.1007/s13280-015-0705-1

Ash, James. Rob Kitchin, and Agnieszka Leszczynski. 2018. "Digital Turn, Digital Geographies?" *Progress in Human Geography* 42: 25–43. https://doi.org/10.1177/0309132516664800

Baker, John Alec. 1967. *The Peregrine*. London: William Collins.

Banks, Alexander N., Humphrey Q.P. Crick, Rachel Coombes, Stuart Benn, Derek A. Ratcliffe and Elizabeth M. Humphreys. (2010) "The Breeding Status of Peregrine Falcons *Falco peregrinus* in the UK and Isle of Man in 2002." *Bird Study* 57: 421–436. https://doi.org/10.1080/00063657.2010.511148

Barrow, Mark. 2009. *Nature's Ghosts: Confronting Extinction from the Age of Jefferson to the Age of Ecology*. Chicago: University of Chicago Press.

BirdLife International. 2022. "Species Factsheet: *Falco peregrinus*." Accessed February 26, 2022. http://www.birdlife.org

Brighton, Caroline H., Adrian L.R. Thomas, and Graham K. Taylor. 2017. "Terminal Attack Trajectories of Peregrine Falcons are Described by the Proportional Navigation Guidance Law of Missiles." *Proceedings of the National Academy of Sciences* 114: 13495–13500. https://doi.org/10.1073/pnas.1714532114

Campanella, Thomas J. 2004. "Webcameras and the Telepresent landscape." In *The Cybercities Reader*, edited by Steve Graham, 57–63. London: Routledge.

Carson, Rachel. 1962. *Silent Spring*. Boston: Houghton Mifflin.

Chambers, Charlotte N.L. 2007. "'Well its Remote, I Suppose, Innit?' The Relational Politics of Bird-Watching through the CCTV Lens." *Scottish Geographical Journal* 123: 122–134. https://doi.org/10.1080/14702540701624568

Chichester Peregrines. 2021. "Thank you AG's Antics this PM….!!" *Chichester Peregrine Blog*, Accessed March 15, 2022. https://chichesterperegrinesblog.co.uk/2021/06/17/thank-you-ags-antics-this-pm/

Crick, Humphrey Q.P. and Derek A. Ratcliffe. 1995. "The Peregrine *Falco peregrinus* Breeding Population of the United Kingdom in 1991." *Bird Study* 42: 1–19. https://doi.org/10.1080/00063659509477143

Dee, Tim. 2009. *Year on the Wing: Four Seasons in a Life with Birds*. New York: Free Press.

Deleuze, Gilles and Guattari, Félix. 1987. *A Thousand Plateaus: Capitalism and Schizophrenia*. Minneapolis: University of Minnesota Press.

Despret, Vinciane. 2021. *Living as a Bird*. Cambridge: Polity Press.

Dixon, Nick. 2000. "A New Era for Peregrines: Buildings, Bridges, and Pylons as Nest Sites." *BTO News* 229: 10–11.

Donázar, José A., Ainara Cortés-Avizanda, Juan A. Fargallo, Antoni Margalida, Marcos Moleón, Zebensui Morales-Reyes, Rubén Moreno-Opo, Juan M. Pérez-García, José A. Sánchez-Zapata, Iñigo Zuberogoitia, and David Serrano. 2016. "Roles of Raptors in a Changing World: From Flagships to Providers of Key Ecosystem Services." *Ardeola* 63: 181–234. https://doi.org/10.13157/arla.63.1.2016.rp8

Drewitt, ed. 2014. *Urban Peregrines*. Exeter: Pelagic Publishing.

Drewitt, Edward J.A. and Nick Dixon. 2008. "Diet and Prey Selection of Urban-Dwelling Peregrine Falcons in Southwest England." *British Birds* 101: 58–67.

Gainzarain, José A., R. Arambarri, and A.F. Rodríguez. 2000. "Breeding Density, Habitat Selection and Reproductive Rates of the Peregrine Falcon *Falco peregrinus* in Álava (Northern Spain)." *Bird Study* 47: 225–231. https://doi.org/10.1080/00063650009461177

Garlick, Ben. 2019. "Cultural Geographies of Extinction: Animal Culture among Scottish Ospreys." *Transactions of the Institute of British Geographers* 44 (2): 226–241. https://doi.org/10.1111/tran.12268

Goode, David. 2014. *Nature in Towns and Cities*. London: William Collins, New Naturalist Library.

Greenwood, Jeremy. 2021. "BB Eye: It Was Not DDT." *British Birds* 114 (5): 248–250. https://britishbirds.co.uk/content/it-was-not-ddt

Heise, Ursula. 2016. *Imagining Extinction: The Cultural Meanings of Endangered Species*. Chicago: Chicago University Press.

Hennen, Mary and Peggy Macnamara. 2017. *The Peregrine Returns: The Art and Architecture of an Urban Raptor Recovery*. Chicago: The University of Chicago Press.

Hinchliffe, Steve, Matthew B. Kearnes, Monica Degen, M., and Sarah Whatmore. 2005. "Urban Wild Things: A Cosmopolitical Experiment." *Environment and Planning D: Society and Space* 23: 643–658. https://doi.org/10.1068/d351t

Hinchliffe, Steve and Sarah Whatmore. 2006. "Living Cities: Towards a Politics of Conviviality." *Science as Culture* 15: 123–138.

Humphreys, Liz, Chris V. Wernham, and Humphrey Crick. 2007. *Raptor Species Conservation Frameworks: The Peregrine Conservation Framework Project Progress Report — Phase I*. Stirling: British Trust for Ornithology.

Jasanoff, Sheila. 2004. "Ordering Knowledge, Ordering Society." In *States of Knowledge: The Co-Production of Science and Social Order*, edited by Sheila Jasanoff, 13–45. London: Routledge.

Jenkins, Andrew R. 2000. "Hunting Mode and Success of African Peregrines *Falco peregrinus minor,* Does Nesting Habitat Quality Affect Foraging Efficiency?" *Ibis* 142: 235–246. https://doi.org/10.1111/j.1474-919X.2000.tb04863.x

Jørgensen, Finn Arne. 2014. "The Armchair Traveler's Guide to Digital Environmental Humanities." *Environmental Humanities* 4: 95–112. https://doi.org/10.1215/22011919-3614944

Kamphof, Ike. 2011. "Webcams to Save Nature: Online Space as Affective and Ethical Space." *Foundations of Science* 16: 259–274. https://doi.org/10.1007/s10699-010-9194-7

Kettel, Esther F. Louise K. Gentle, and Richard W. Yarnell. 2016. "Evidence of an Urban Peregrine Falcon (*Falco peregrinus*) Feeding Young at Night." *Journal of Raptor Research* 50: 321–323. https://doi.org/10.3356/JRR-16-13.1

Kettel, Esther F. Louise K. Gentle, Richard W. Yarnell, and John L. Quinn. 2019. "Breeding Performance of an Apex Predator, the Peregrine Falcon, Across Urban and Rural Landscapes." *Urban Ecosystems* 22: 117–125. https://doi.org/10.1007/s11252-018-0799-x

Koskela, Hille. 2003. "'Cam Era'— The Contemporary Urban Panopticon." *Surveillance and Society* 1: 292–313. https://doi.org/10.24908/ss.v1i3.3342

Koskela, Hille. 2004. "Webcams, TV shows and Mobile phones: Empowering Exhibitionism." *Surveillance and Society* 2: 199–215. https://doi.org/10.24908/ss.v2i2/3.3374

Lorimer, Jamie. 2007. "Nonhuman Charisma." *Environment and Planning D: Society and Space* 25: 911–932. https://doi.org/10.1068/d71j

Lorimer, Jamie. 2008a. "Living Roofs and Brownfield Wildlife: Towards a Fluid Biogeography of UK Nature Conservation." *Environment and Planning A*, 40: 2042–2060. https://doi.org/10.1068/a39261

Lorimer, Jamie. 2008b. "Counting Corncrakes: The Affective Science of the UK Corncrake Census." *Social Studies of Science* 38: 377–405. https://doi.org/10.1177/0306312707084396

Lorimer, Jamie. 2020. *The Probiotic Planet: Using Life to Manage Life.* Minneapolis: University of Minnesota Press.

Lorimer, Jamie, Timothy Hodgetts, and Maan Barua. 2019. "Animals' Atmospheres." *Progress in Human Geography* 43: 26–45. https://doi.org/10.1177/0309132517731254

Macdonald, Helen. 2006. *Falcon.* London: Reaktion.

Macfarlane, Robert. 2017. "Afterword." In *The Peregrine,* edited by J. A. Baker, 193–210. London: William Collins.

MacKenzie, John. 1988. *The Empire of Nature: Hunting, Conservation and British Imperialism.* Manchester: Manchester University Press.

Mak, Brandon, Robert A. Francis, and Michael A. Chadwick. 2021a. "Breeding Habitat Selection of Urban Peregrine Falcons (*Falco peregrinus*) in London." *Journal of Urban Ecology* 7 (1). https://doi.org/10.1093/jue/juab017

Mak, Brandon, Robert A. Francis, and Michael A. Chadwick. 2021b. "Living in the Concrete Jungle: A Review and Socio-ecological Perspective of Urban Raptor Habitat Quality in Europe." *Urban Ecosystems* 24: 1179–1199. https://doi.org/10.1007/s11252-021-01106-6

Marris, Emma. 2010. *Rambunctious Garden: Saving Nature in a Post-Wild World.* New York: Barnes and Noble.

McCorristine, Shane and William M. Adams. 2020. "Ghost Species: Spectral Geographies of Biodiversity Conservation." *Cultural Geographies* 27: 101–11. https://doi.org/10.1177/1474474019871645

McLean, Jessica. 2020. *Changing Digital Geographies: Technologies, Environments, and People.* London: Palgrave Macmillan.

Mellanby, Kenneth. 1967. *Pesticides and Pollution.* London: Collins.

Molard, Laurent, Marc Kéry, and Clayton M. White. 2007. "Estimating the Resident Population Size of Peregrine Falcon *Falco peregrinus* in Peninuslar Malaysia." *Forktail* 23: 87–91.

Moss, Timothy, Friederike Voigt, and Sören Becker. 2021. "Digital Urban Nature." *City* 25: 255–276. https://doi.org/10.1080/13604813.2021.1935513

Newton, Ian. 2015. "Pesticides and Birds of Prey: The Breakthrough." In *Nature's Conscience: The Life and Legacy of Derek Ratcliffe*, edited by Desmond Thompson, Hilary H. Birks, and John Birks, 281–299. Kings Lynn: Langford Press.

Oliver, Catherine. 2021. "OurChickenlife: Byproductive Labour in the Digital Flocl." *Digital Ecologies*, May 6, 2021. Accessed June 3, 2022. http://www.digicologies.com/2021/05/06/catherine-oliver/

Pagel, Joel E., Clifford M. Anderson, Douglas A. Bell, Edward Deal, Lloyd Kiff, F. Arthur McMorris, Patrick T. Redig, and Robert Sallinger. 2018. "Peregrine Falcons: The Neighbours Upstairs." In *Urban Raptors: Ecology and Conservation of Birds of Prey in Cities*, edited by Clint W. Boal and Cheryl R. Dykstra, 180–195. Washington, DC: Island Press.

Pitas, John-Henry. 2022. "Deathly Storytelling in the Ecological City: How Pigeons Became Falcon Food in Baltimore, Maryland." *Social & Cultural Geography* 23 (1): 29–46. https://doi.org/10.1080/14649365.2021.1950822

Ratcliffe, Derek A. 1963. "The Status of the Peregrine in Great Britain." *Bird Study* 10: 56–90. https://doi.org/10.1080/00063656309476042

Ratcliffe, Derek A. 1970. "Changes Attributable to Pesticides in Egg Breakage Frequency and Eggshell Thickness in Some British Birds." *Journal of Applied Ecology* 7: 67–115. https://doi.org/10.2307/2401613

Ratcliffe, Derek A. 1972. "The Peregrine population of Great Britain in 1971." *Bird Study* 19: 117–156. https://doi.org/10.1080/00063657209476336

Ratcliffe, Derek A. 1980. *The Peregrine Falcon*. Berkhamsted: T. & A. D. Poyser.

Ratcliffe, Derek A. 1984. "The Peregrine Breeding Population of the United Kingdom in 1981." *Bird Study*, 31: 1–18. https://doi.org/10.1080/00063658409476809

Ratcliffe, Derek A. 1993. *The Peregrine Falcon* (2nd ed.). Berkhamsted: T and A.D. Poyser.

Rose, Gillian. 2015. "Rethinking the Geographies of Cultural 'Objects' Through Digital Technologies: Interface, Network and Friction." *Progress in Human Geography* 40: 334–351. https://doi.org/10.1177/0309132515580493

RSPB. 2021 "Peregrines: Population Numbers and Trends." Royal Society for the Protection of Birds. Accessed November 1, 2021. https://www.rspb.org.uk/birds-and-wildlife/wildlife-guides/bird-a-z/peregrine/population-numbers-and-trends/

Schilthuizen, Menno. 2018. *Darwin Comes to Town: How the Urban Jungle Drives Evolution*. New York: Picador.

Schroer, Sara Asu. 2021. "Caring for Falcons in a Time of Extinction." *Theorizing the Contemporary, Cultural Anthropology Fieldsights*, January 26. https://culanth.org/fieldsights/caring-for-falcons-in-a-time-of-extinction

Scoville, Caleb, Melissa Chapman, Razvan Amironesei, and Carl Boettiger. 2021. "Algorithmic Conservation in a Changing Climate." *Current Opinion in Environmental Sustainability* 51: 30–35. https://doi.org/10.1016/j.cosust.2021.01.009

Searle, Adam. 2020. "Absence." *Environmental Humanities* 12: 167–172. https://doi.org/10.1215/22011919-8142253

Searle, Adam. 2021. "Hunting Ghosts: on Spectacles of Spectrality and the Trophy Animal." *Cultural Geographies* 28: 513–530. https://doi.org/10.1177/1474474020987250

Searle, Adam. 2022. "Spectral Ecologies: De/extinction in the Pyrenees." *Transactions of the Institute of British Geographers* 47: 167–183. https://doi.org/10.1111/tran.12478

Searle, Adam, Jonathon Turnbull, and William M. Adams. 2023. "The Digital Peregrine: A Technonatural History of A Cosmopolitan Raptor." *Transactions of the Institute of British Geographers*. 48: 195-212. https://doi.org/10.1111/tran.12566

Sheail, John. 1976. *Nature in Trust: The History of Nature Conservation in Britain*. Glasgow: Blackie.

Sheail, John. 1985. *Pesticides and Nature Conservation: The British Experience 1950–1975*. Oxford: Clarendon Press.

Stefik, Mark J., ed. 1997. *Internet Dreams: Archetypes, Myths, and Metaphors*. Cambridge: MIT Press.

Stirling-Aird, Patrick. 2015. *Peregrine Falcon*. London: Bloomsbury.

Time, Bjarne Emil. 2016. "Hunting Activity by Urban Peregrine Falcons (*Falco peregrinus*) During Autumn and Winter in South-west Norway." *Ornis Norvegica* 39: 39–44. https://doi.org/10.15845/on.v39i0.1048

Tucker, John. 1998. "The Peregrine Falcon in Shropshire: Whatever Next?" *British Wildlife* 9: 227–231.

van der Wal, René, and Koen Arts. 2015. "Digital Conservation: An Introduction." *Ambio*, 44: 517–521. https://doi.org/10.1007/s13280-015-0701-5

van Dooren, Thom. 2014. *Flight Ways: Life and Loss at the Edge of Extinction*. New York: Columbia University Press.

Verma, Audrey, René van der Wal, and Anke Fischer. 2015. "Microscope and Spectacle: On the Complexities of Using New Visual Technologies to Communicate about Wildlife Conservation." *Ambio* 44: 648–660.

von Essen, Erica. 2021. "Digital Biosurveillance: How Digital Ecology is About Capturing Movement." *Digital Ecologies*, Jun 7, 2021. Accessed June 2, 2022. http://www.digicologies.com/2021/06/07/erica-von-essen/

von Essen, Erica, Jonathon Turnbull, Adam Searle, Finn Arne Jørgensen, Tim R. Hofmeester, and René van der Wal. 2021. "Wildlife in the Digital Anthropocene: Examining Human-animal Relations through Surveillance Technologies." *Environment and Planning E: Nature and Space*. Online: https://doi.org/10.1177/25148486211061704

Washburn, Brian E. 2018. "Human-raptor Conflicts in Urban Settings." In *Urban Raptors: Ecology and Conservation of Birds of Prey in Cities*, edited by Clint W. Boal and Cheryl R. Dykstra, 214–228. Washington, DC: Island Press.

White, Clayton M, Sarah A. Sonsthagen, George K. Sage, Clifford Anderson, and Sandra L. Talbot. 2013. "Genetic Relationships among Some Subspecies of the Peregrine Falcon (*Falco peregrinus* L.), Inferred from Mitochondrial DNA Control-Region Sequences." *Auk* 130: 78–87.

Wilson, M.W., E. Balmer, K. Jones, V. A. King, D. Raw, C. J. Rollie, E. Rooney, M. Ruddock, G. D. Smith, A. Stevenson, P. K. Stirling-Aird, C. V. Wernham, J. M. Weston, and D. G. Noble. 2018. "The Breeding Population of Peregrine Falcon *Falco peregrinus* in the United Kingdom, Isle of Man and Channel Islands in 2014." *Bird Study* 65: 1–19.

Wrigley, Charlotte. 2020. "Nine Lives Down: Love, Loss, and Longing in Scottish Wildcat Conservation." *Environmental Humanities* 12: 346–369. https://doi.org/10.1215/22011919-8142396

Yusoff, Kathryn. 2012. "Aesthetics of Loss: Biodiversity, Banal Violence and Biotic Subjects." *Transactions of the Institute of British Geographers* 37 (4): 578–592. http://www.jstor.org/stable/41678656

INDEX

Pages in *italics* refer figures, and pages followed by n refer notes.

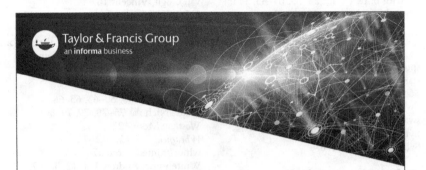

Printed in the United States
by Baker & Taylor Publisher Services